NB

Collins Revision

GCSE Foundation Maths

Revision Guide

FOR EDEXCEL A
FOR AQA A
FOR AQA B

Keith Gordon

Contents

Algebra

Statistical representation

Statistics

- Statistics is concerned with the collection and organisation of **data**.
- It is usual to use a **data collection sheet** or **tally chart** for collecting data.
- Data is recorded by means of **tally marks**, which are added to give a **frequency**.
- Data can be collected in a **frequency table**.

This tally chart shows the results when Alf threw two coins together several times.

Outcome	Tally	Frequency
2 tails	IIII IIII IIII	14
1 tail, 1 head	IIII IIII IIII IIII IIII III	
2 heads	IIII IIII IIII III	

Collecting data

- Data can be collected by three methods.
  - **Taking a sample**, for example, to find which 'soaps' students watch, ask a random selection of 50 students.
  - **Observation**, for example, to find out how many vehicles use a road in a day, count the vehicles that pass during an hour.
  - **Experiment**, for example, to see if a home-made spinner is biased, throw it about 100 times and record the outcomes.

Top Tip!

When collecting or tallying, always double-check as it is easy to miss a piece of data.

Pictograms

- A pictogram is a way of showing data in a diagrammatic form that uses **symbols** or **pictures** to show **frequencies**.
- Every pictogram has a **title** and a **key** that shows how many items are represented by each symbol.

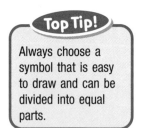

Top Tip!

Always choose a symbol that is easy to draw and can be divided into equal parts.

Average daily number of hours of sunshine

Month	Hours of sunshine	Total
June		11
July		
August		13

Key: ☼ represents 4 hours

Questions

Grade G

1 a Copy the tally chart, above, and complete the frequency column.

  b Which outcome is most likely?

  c How many times altogether did Alf throw the two coins?

Grade G

2 The pictogram, above, shows the average daily hours of sunshine for a town in Greece during June, July and August.

  a What was the daily average for July?

  b Complete the pictogram for August.

Remember: You must revise all content from Grade G to the level that you are currently working at.

F

Bar charts

- A **bar chart** is made up of bars or blocks of the same width, drawn horizontally or vertically on an axis.

- The **heights** or **lengths** of the bars always represent **frequencies**.

- The bars are separated by small gaps to make the chart easier to read.

- Both **axes** should be **labelled**.

- The **bar chart** should be **labelled** with a **title**.

- **A dual bar chart** can be used to compare two sets of data.

This graph shows sales of two newpapers over a week.

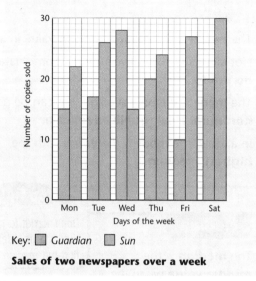

Key: ☐ *Guardian* ☐ *Sun*

Sales of two newspapers over a week

E-D

Line graphs

- **Line graphs** are used to show how data changes over a period of time.

- Line graphs can be used to show **trends**, for example, how the average daily temperature changes over the year.

- **Data points** on line graphs can be joined by **lines**.
  - When the lines join points that show **continuous data**, for example, joining points showing the height of a plant each day over a week, they are drawn as **solid lines**. This is because the lines can be used to estimate intermediate values.
  - When the lines join points that *do not* show continuous data, for example, the daily takings of a corner shop, they are drawn as **dotted lines**. This is because the lines cannot be used to estimate values, they just show the **trend**.

This graph compares temperatures.

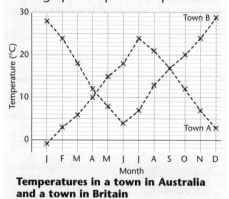

Temperatures in a town in Australia and a town in Britain

Top Tip!

The largest increase or decrease on a line graph is at the steepest part of the graph.

Questions

Grade F

1 Refer to the dual bar chart, above.

  a On which two days were the same number of copies of the *Guardian* sold?

  b How many copies of the *Sun* were sold altogether over the week?

  c On which day was there the greatest difference between the numbers of copies sold of the *Guardian* and the *Sun*?

  d On one day the *Guardian* had a 'half-price' offer. Which day was this? Explain your answer.

Grade E

2 Refer to the line graph, above.

  a Which town is hotter, on average? Give a reason for your answer.

  b Which town is in Australia? Give a reason for your answer.

  c In which month was the average temperature the same in both towns?

  d Is it true that the average temperature was the same in both towns on a day in early April?

Averages

- An **average** is a single value that gives a **representative value** for a set of data. There are three types of average: **mode**, **median** and **mean**.

Mode

- The **mode** is the **most common** value in a set of data.
- Not all sets of data have a mode. Some data sets have **no mode**.
- The mode is the only average that can be used for **non-numerical (qualitative)** data.
- In a table, the mode is the **value** with the **highest frequency**.

These are the numbers of eggs in 20 sparrows' nests.

2, 4, 1, 3, 5, 1, 3, 1, 4, 5,
1, 2, 3, 3, 1, 5, 2, 1, 3, 2

The **modal number** of eggs per nest is 1.

Median

Top Tip!

Don't forget to put the data in order, to find the median.

- The **median** is the **middle value** when the data items are arranged in order.
- If there are n pieces of data the median is the $\frac{n+1}{2}$ **th** number in the list.
- If the number of items of data is **odd**, the middle value will be one of the data items.
- If the number of items of data is **even**, the middle value will be halfway between two of the data items.
- To find the median in a **frequency table**, add up the frequencies in the table until the total passes half the number of data items in the whole set.

This frequency table has been used to record results of a survey.

Value	3	4	5	6	7	8
Frequency	7	8	9	12	3	1
Total frequency	7	15	24			

There are 40 values so the median is the $\frac{40+1}{2} = 20\frac{1}{2}$th value.
Add up the frequencies until you find where this value will be.
The median must be in the third column, so the median is 5.

Mean

- The **mean** is defined as $\text{mean} = \dfrac{\text{sum of all values}}{\text{number of values}}$
- When people talk about 'the average' they are usually referring to the mean.
- The advantage of using the mean is that it **takes all values into account**.

These are the numbers of eggs in 20 sparrows' nests.

2, 4, 1, 3, 5, 1, 3, 1, 4, 5,
1, 2, 3, 3, 1, 5, 2, 1, 3, 2

The **mean number** of eggs per nest is 52 ÷ 20 = 2.6.

Questions

Grade G

1 **a** Find the mode for each set of data.
   i 4, 5, 8, 4, 3, 5, 6, 4, 5, 7, 9, 5, 3, 8
   ii Y, B, Y, R, G, Y, R, G, Y, Y, R, B, B, Y, R
   b Why is the modal number of eggs in sparrows' nests, above, not a good average?

Grade F

2 Find the median for each set of data.
   a 8, 7, 3, 2, 10, 8, 6, 5, 9
   b 5, 13, 8, 7, 11, 6, 12, 9, 15, 4, 3, 10

Grade F

3 Use a calculator to find the mean of each data set.
   a 44, 66, 99, 44, 34, 66, 83, 70, 45, 76, 77
   b 34.5, 44.8, 29.3, 27.2, 34.1, 39.0, 30.4, 40.7

Remember: You must revise all content from Grade G to the level that you are currently working at.

G-E

Range

- The **range** of a set of data is the **highest value minus the lowest value**.

  Range = highest value − lowest value

- The range is used to measure the **spread of data**.

- The range is used to comment on the **consistency of the data**. A smaller range indicates more consistent data.

Top Tip!

Remember! The range is *not* an average.

F-C

Which average to use

- The average must be truly **representative** of the data, so the average used must be **appropriate** for the set of data.

- The table below shows the advantages and disadvantages of each average.

	Mode	Median	Mean
Advantages	• Easy to find • Not affected by extreme values	• Easy to find for ungrouped data • Not affected by extreme values	• Uses all values • Original total can be calculated
Disadvantages	• Does not use all values • May be at one extreme • May not exist	• Does not use all values • Raw data has to be put in order • Hard to find from table	• Most difficult to calculate • Extreme values can distort it
Use for	• Non-numerical data • Data sets in which a large number of values are the same	• Data with extreme values	• Data with values that are spread in a balanced way • Data in which all values are relevant, for example, cricket score averages

Questions

Grade E

1 Find the range of each of these sets of data.
  a 4, 9, 8, 5, 6, 10, 11, 7, 8, 5
  b −4, 8, 5, −1, 0, 0, 2, 4, 3, −2, −3, 4

Grade E

2 For a darts match, the captain has to choose between Don and Dan to play a round. During the practice, the captain records the scores they both make with three darts on five throws. These are:
  Dan 120, 34, 61, 145, 20   **Don** 68, 89, 80, 72, 71
  a Work out the mean for each player.

  b Work out the range for each player.
  c Who should the captain pick? Explain why.

Grade C

3 After a maths test, the class were told the mean, mode and median marks.
  Three students made these statements.
  Asaf: 'I was in the top half of the class.'
  Brian: 'I can't really tell how well I have done.'
  Clarrie: 'I have done really well compared with the rest of the class.'
  Which averages did they use to make these statements?

Arranging data

Frequency tables

- When you have to represent a lot of data, use a **frequency table**.

This table shows how many times students in a form were late in a week.

Number of times late	0	1	2	3	4	5
Frequency	11	8	3	3	3	2

Top Tip!

The mode is the value that has the highest frequency; it is not the actual frequency.

D–C

- The **mode** is the data value with the **highest frequency**.

- Find the **median** by adding up the frequencies of the data items, in order, until the halfway point of all the data in the set is passed.

- Find the **mean** by multiplying the value of each data item by its frequency, adding the totals, then dividing by the total of all the frequencies.

To find the average number of times students were late, work out the total number of times students were late: $0 \times 11 + 1 \times 8 + 2 \times 3 + 3 \times 3 + 4 \times 3 + 5 \times 2$ and work out the total frequency: $11 + 8 + 3 + 3 + 3 + 2$, then divide the first total by the second.

Grouped data

- A wide range of data, with lots of values, may have too many entries for a frequency table, so use a **grouped frequency table**.

- Grouped data can be shown by a **frequency polygon**.

- For a frequency polygon, plot the **midpoint** of each group against the **frequency**.

- In a grouped frequency table, data is **recorded in groups** such as $10 < x \leq 20$.

- $10 < x \leq 20$ means values between 10 and 20, not including 10 but including 20.

- The **modal class** is the group with the greatest frequency. It is not possible to identify the actual mode.

- The **median** cannot be found from a grouped table.

- Calculate an **estimate of the mean** by adding the midpoints multiplied by the frequencies and dividing the result by the total frequency.

This table shows the marks in a mathematics examination for 50 students. The frequency polygon shows the data.

Marks, x	Frequency, f
$0 < x \leq 10$	4
$10 < x \leq 20$	9
$20 < x \leq 30$	17
$30 < x \leq 40$	13
$40 < x \leq 50$	7

C

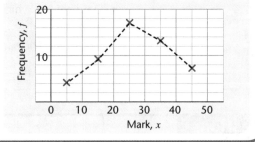

Questions

Grade D

1 Study the table showing how often students were late in a week.

  a How many students were there in the form?

  b What is the modal number of times they were late?

  c **i** What is the total number of 'lates'?

    ii What is the mean number of 'lates' per day?

Grade C

2 The marks for 50 students in a mathematics examination are shown in the table, above.

  a What is the modal class?

  b **i** What is the total of the 'midpoints times frequencies'?

    ii What is the estimated mean mark for the form?

Stem-and-leaf diagrams

D-C

- When data is first recorded, it is called **raw data** and is **unordered**.

  The ages of the first 15 people to use a shop in the morning were:

  18, 26, 32, 29, 31, 57, 42, 16, 23, 42, 30, 19, 42, 35, 38

- Unordered data can be put into order to make it easier to read and understand. This is called **ordered** data.

- A **stem-and-leaf diagram** is a way of showing ordered data.

  This stem-and-leaf diagram shows the marks scored by students in a test.

1	3 4 5 6
2	0 1 1 3 4 5 8
3	1 2 5 5 5 9
4	2 2 9

  Key 1 | 3 represents 13 marks

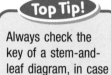
Top Tip!
Always check the key of a stem-and-leaf diagram, in case the 'stem' is not tens.

- The **stem** is the number on the left of the vertical line. In this case, it represents the tens.

- The **leaves** are the numbers on the right of the vertical line. In this case, they represent the units.

- A stem-and-leaf diagram must have a **key**, which shows what the stem and the leaves stand for. The stem is *not always tens*.

- In the diagram above the first six entries are 13, 14, 15, 16, 20 and 21.

- Find the **mode** by finding the most common entry.

- Find the **median** by counting from the start to the middle value.

- Find the **range** by subtracting the lowest value from the highest value.

- Find the **mean** by adding all the values and dividing by the total number of values in the table.

Questions

Grade D

1 The ages of the first 15 people to use a shop in the morning are shown, above.

  a Rearrange the raw data into an ordered list and draw a stem-and-leaf diagram.

  b What is the modal age?

  c Why is the mode not a good average to use for this data?

  d What is the median age?

  e What is the range of the ages?

  f What is the mean age?

Grade D

2 The marks for some students in a test are shown in the stem-and-leaf diagram, above.

  a How many students took the test?

  b i What is the lowest mark in the test?

  ii What is the highest mark in the test?

  iii What is the range of the marks?

  c What is the modal mark?

  d What is the median mark?

  e i What is the total of all the marks?

  ii What is the mean mark?

Probability

The probability scale

- People use everyday words such as '**chance**', '**likelihood**' and '**risk**' to assess whether something will happen.

  'What is the chance of rain today?' or 'Shall we risk not taking an umbrella?'

- **Probability** is the mathematical method of assessing chance, likelihood or risk.

- Probability gives a value to the **outcome** of an **event**.

  The probability of throwing a six with a dice is one-sixth.

- The **probability scale** runs from **0 to 1**.

- Terms such as '**very likely**' and '**evens**' are used to associate probabilities with approximate positions on the scale.

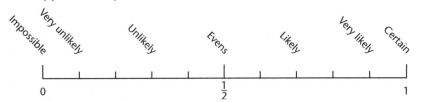

- The probability of an **impossible** event is 0.   The probability that pigs will fly is 0.

- The probability of a **certain** event is 1.   The probability that the sun will rise tomorrow is 1.

> **Top Tip!**
>
> Probabilities may be given as fractions or percentages but they can never be less than zero or greater than 1.

Calculating probabilities

- Whenever anyone does something, such as take a card from a pack of cards or buy a raffle ticket, there are a number of **possible outcomes**.

  There are 52 possible outcomes of picking a card from a pack of cards.

- There may be **more than one way** of any particular outcome of an event occurring.

  There are four possible ways of picking a king from a pack of cards.

- The probability of an outcome of an event is defined as:

  $$P(\text{outcome}) = \frac{\text{number of ways the outcome can occur}}{\text{total number of possible outcomes}} \qquad P(\text{king}) = \frac{4}{52} = \frac{1}{13}$$

- **At random** means 'without looking' or 'not knowing the outcome in advance'.

Questions

Grade G

1 a State whether each event is impossible, very unlikely, unlikely, evens, likely, very likely or certain.

   i   Christmas Day being on 25 December.

   ii Scoring an even number when a regular dice is thrown.

   iii Someone in the class having a mobile phone.

   iv A dog talking.

  b Draw a probability scale and mark an arrow for the approximate probability of each outcome.

   i   The next person to walk through the door will be female.

   ii The person sitting next to you in mathematics is under 18 years of age.

   iii Someone in the class will have a haircut today.

Grade F

2 What is the probability of each of these events?

  a Throwing a 6 with a dice.

  b Picking a picture card from a pack of cards.

  c Tossing a tail with a coin.

(**Remember:**) You must revise all content from Grade G to the level that you are currently working at.

E

Probability of 'not' an event

- To find the probability that an event **does not happen**, use the rule:
  P(event) + P(not event) = 1 or **P(not event) = 1 – P(event)**

  If the probability that a student picked at random from a class is a girl is $\frac{5}{11}$, then the probability that a student picked at random is not a girl (is a boy) is $1 - \frac{5}{11} = \frac{6}{11}$.

Top Tip!

When subtracting a fraction from 1, just find the difference between the numerator and the denominator for the new numerator, for example, $1 - \frac{3}{8} = \frac{(8-3)}{8} = \frac{5}{8}$.

D-C

Addition rule for mutually exclusive outcomes

- **Mutually exclusive** outcomes or events cannot occur at the same time.

- For the probability of either of two mutually exclusive events, **add** the probabilities of the separate events: **P(A or B) = P(A) + P(B)**

Top Tip!

This rule only works with mutually exclusive events.

  Picking a jack and a king from a pack of cards are mutually excusive events.

  The probability of picking a jack or a king is: $\frac{4}{52} + \frac{4}{52} = \frac{8}{52} = \frac{2}{13}$

  The probability of picking a king or a red card is not $\frac{4}{52} + \frac{26}{52}$ because there are two red kings.

C

Relative frequency

- The probability of an event can be found by three methods.

  - **Theoretical probability** for **equally likely outcomes**, such as drawing from a pack of cards.

  - **Experimental probability**, caculating relative frequency, for example, from a large number of trial spins of a home-made spinner.

  - **Historical data**: looking up past records, for example, for an earthquake in Japan.

- To calculate the relative frequency of an event, **divide** the **number of times the event occurred** during the experiment by the **total number of trials**.

- The **more trials** that are done, the nearer the experimental probability will be to the true probability.

Top Tip!

If you are asked who has the best set of results, always say the person with the most trials.

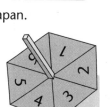

Questions

(Grade E)

1 The probability that someone picked at random is left-handed is $\frac{3}{10}$. What is the probability that someone picked at random is not left-handed?

(Grade E)

2 What is the probability that a card picked randomly from a pack is:

a an ace

b a king

c an ace or a king?

(Grade C)

3 These are the results when three students tested the spinner, shown above.

Student	Ali	Barry	Clarrie
Number of throws	20	60	240
Number of 4s	5	13	45

a For each student, calculate the relative frequency of a 4. Give your answers to 2 decimal places.

b Which student has the most reliable estimate of the actual probability of a 4? Explain why.

c If the spinner was fair, how many times would you expect it to land on 4 in 240 spins?

E

Combined events

- When two events occur together they are known as **combined events**.

- The outcomes of the combined events can be shown as a **list**.

  If two coins are thrown together, the possible outcomes are:
  (head, head), (head, tail), (tail, head) and (tail, tail).

- Another method for showing the outcomes of a combined event is to use a **sample space diagram**.

If two dice are thrown, the possible outcomes can be shown like this.

6	(1, 6)	(2, 6)	(3, 6)	(4, 6)	(5, 6)	(6, 6)
5	(1, 5)	(2, 5)	(3, 5)	(4, 5)	(5, 5)	(6, 5)
4	(1, 4)	(2, 4)	(3, 4)	(4, 4)	(5, 4)	(6, 4)
Score on red dice 3	(1, 3)	(2, 3)	(3, 3)	(4, 3)	(5, 3)	(6, 3)
2	(1, 2)	(2, 2)	(3, 2)	(4, 2)	(5, 2)	(6, 2)
1	(1, 1)	(2, 1)	(3, 1)	(4, 1)	(5, 1)	(6, 1)
	1	2	3	4	5	6

Score on blue dice

If two dice are thrown together and the scores are added, the possible outcomes can be shown like this.

6	7	8	9	10	11	12
5	6	7	8	9	10	11
4	5	6	7	8	9	10
Score on red dice 3	4	5	6	7	8	9
2	3	4	5	6	7	8
1	2	3	4	5	6	7
	1	2	3	4	5	6

Score on blue dice

Questions

Grade E

1 a When two dice are thrown together, how many possible outcomes are there?

b Refer to the top sample space diagram for throwing two dice, above.

   i What is the probability of throwing a double with two dice?

   ii What is the probability that the difference between the scores on the two dice is 4?

c Refer to the bottom sample space diagram for throwing two dice, above.

   i What is the probability of throwing a score of 5 with two dice?

   ii What is the probability of throwing a score greater than 9 with two dice?

   iii What is the most likely score with two dice?

Remember: You must revise all content from Grade G to the level that you are currently working at.

D

Expectation

- When the probability of an event is known we can **predict** how many times the event is likely to happen in a given number of trials.

- This is the **expectation**. It is *not* what is going to happen.

  If a coin is tossed 1000 times we would expect 500 heads and 500 tails. It is very unlikely that we would actually get this result in real life.

- The **expected number** is calculated as: **Expected number = P(event) × total trials**

G-C

Two-way tables

- A two-way table is a table that links two variables.

  This table shows the languages taken by the boys and girls in Form 9Q.

	French	Spanish
Boys	7	5
Girls	6	12

- One of the variables is shown by the rows of the table.

- One of the variables is shown by the columns of the table.

  This table shows the nationalities of people on a jet plane and the types of ticket they have.

	First class	Business class	Economy
American	6	8	51
British	3	5	73
French	0	4	34
German	1	3	12

Questions

Grade D

1 A bag contains 10 counters. Five are red, three are blue and two are white. A counter is taken from the bag at random. The colour is noted and the counter is replaced in the bag. This is repeated 100 times.

a How many times would you expect a red counter to be taken out?

b How many times would you expect a white counter to be taken out?

c How many times would you expect a red or a white counter to be taken out?

Grade D

2 a Refer to the two-way table for Form 9Q, above. Nobody takes two languages.

  i How many boys take Spanish?

  ii How many students are in the form altogether?

  iii How many students take French?

  iv What is the probability that a student picked at random from the form is a girl who takes Spanish?

b Refer to the two-way table for the plane travellers, above.

  i How many travellers were there on the plane?

  ii What percentage of the travellers had first-class tickets?

  iii What percentage of the business-class passengers were American?

Pie charts

Pie charts

- A **pie chart** is another method of representing data.

- Pie charts are used to show the **proportions** between different categories of the data.

- The **angle** of each sector (slice of pie) is **proportional** to the **frequency** of the category it represents.

F-E

This table shows the favourite colours of 20 pupils.

The angle is worked out by multiplying the frequency by 360 divided by the total frequency.

Colour	Frequency	Calculation	Angle
Red	4	4 × 360 ÷ 20	72°
Blue	7	7 × 360 ÷ 20	126°
Green	5	5 × 360 ÷ 20	90°
Yellow	4	4 × 360 ÷ 20	72°
		Total	360°

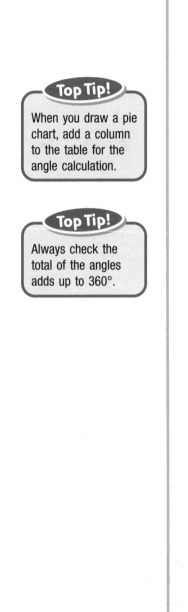

Top Tip!

When you draw a pie chart, add a column to the table for the angle calculation.

Top Tip!

Always check the total of the angles adds up to 360°.

- Pie charts *do not* show individual frequencies, they *only* show proportions.

- Pie charts should always be **labelled**.

This pie chart shows the types of transport used by a group of people going on holiday.

Type of transport

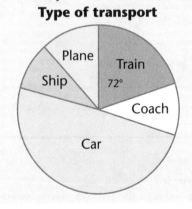

Questions

Grade E

1 Some people were asked what type of transport they mainly used on their holidays. Their replies are shown in the 'Type of transport' pie chart, above.

  a There were 24 people who replied 'Train'.
    How many people were in the survey altogether?

  b There were 11 people who replied 'Ship'.
    What is the angle of the sector representing 'Ship'?

Scatter diagrams

Scatter diagrams

E-C

- A **scatter diagram** (also known as a **scattergraph** or **scattergram**) is a diagram for comparing two variables.
- The variables are plotted as **coordinates**, usually from a table.

Here are 10 students' marks for two tests.

Tables	3	7	8	4	6	3	9	10	8	6
Spelling	4	6	7	5	5	3	10	10	9	7

This is the scatter diagram for these marks.

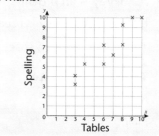

Correlation

E-C

- The scatter diagram will show a **relationship** between the variables if there is one.
- The relationship is described as **correlation** and can be written as a **'real-life' statement**.

For the first diagram: 'The taller people are, the bigger their arm span'.

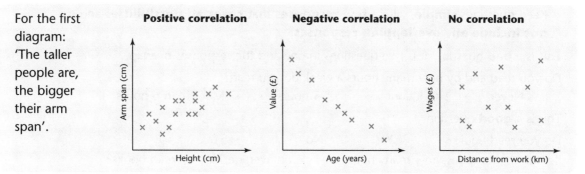

Positive correlation   **Negative correlation**   **No correlation**

Line of best fit

E-C

- A **line of best fit** can be drawn through the data.
- The line of best fit can be used to **predict** the value of one variable when the other is known.

Top Tip!

Draw the line of best fit between the points with about the same number of points on either side of it.

The line of best fit passes through the 'middle' of the data.

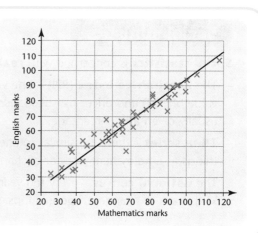

Questions

Grade D

1 a Refer to the scatter diagram of students' scores for tables and spelling tests, above. What type of correlation does the scatter diagram show?

b Describe the relationship in words.

Grade C

2 a Describe, in words, the relationship between the value of a car and the age of the car, shown in the second of the three scatter diagrams, above.

b Describe, in words, the relationship between the wages and the distance travelled to work, shown in the third of the three scatter diagrams, above.

Grade C

3 Refer to the line of best fit on the scatter diagram showing the English and mathematics marks. Estimate the score in the English examination for someone who scored 75 in the mathematics examination.

Surveys

Surveys

- **A survey** is an organised way of finding people's opinions or testing an hypothesis.
- Data from a survey is usually collected on a **data collection sheet**.

  This data collection sheet shows the favourite types of chocolate for 50 students.

Type of chocolate	Tally	Frequency
Milk	ℋℋ ℋℋ ℋℋ ℋℋ ℋℋ III	28
Plain	ℋℋ ℋℋ ℋℋ II	17
Fruit and nut	ℋℋ ℋℋ I	11
White	ℋℋ III	8

- **Questionnaires** are used to collect a lot of data.
- **Questions** on questionnaires should follow some rules.
  - Never ask a **leading question** or a **personal question**.
  - Keep questions **simple**, with a **few responses** that **cover all possibilities** and **do not include any overlapping responses**.

  This is a **bad** question: it is two questions in one and the responses overlap.

  Do you read and by how many hours a week do you read?
  ☐ Never ☐ 0–1 hour ☐ 2–5 hours ☐ More than 6 hours

  This is a **good** question.

  Do you read books? ☐ Yes ☐ No

  If your answer is yes, how many hours a week, on average, do you read books?
  ☐ Up to 2 hours ☐ Between 2 and 4 hours ☐ Over 4 hours

Social statistics

- **Social statistics** is concerned with **real-life statistics**, such as those listed below.
  - The **Retail Price Index** (RPI) – one year is chosen as a reference year (usually given a value of 100) and prices for subsequent years are compared to this, with the change usually given as a percentage.
  - **Time series** – these are similar to line graphs and they plot the changes over time of such things as employment rates or exchange rates between the pound and the dollar.
  - The **national census** – every ten years the government takes a national census so they can keep track of changes in populations. Censuses help governments plan for future changes.

Questions

Grade D

1 Here are two questions used in a survey about recycling. Give a reason why each question is not a good one.
  a Recycling is a waste of time and does not help the environment. Don't you agree?
  ☐ Yes ☐ No
  b How many times a month do you use a bottle bank?
  ☐ Never ☐ 2 times or less
  ☐ More than 4 times

Grade C

2 The euro was introduced in January 2002. Initially the exchange rate between the euro and the pound was €1.65 = £1. In January 2007 the exchange rate was €1.40 = £1. Taking January 2002 as the reference year, with an exchange rate index of 100, what is the exchange rate index for January 2007?

Handling data checklist

I can...

- [ ] draw and read information from bar charts, dual bar charts and pictograms
- [ ] find the mode and median of a list of data
- [ ] understand basic terms such as 'certain', 'impossible', 'likely'

You are working at (Grade G) level.

- [ ] work out the total frequency from a frequency table and compare data in bar charts
- [ ] find the range of a set of data
- [ ] find the mean of a set of data
- [ ] understand that the probability scale runs from 0 to 1
- [ ] calculate the probability of events with equally likely outcomes
- [ ] interpret a simple pie chart

You are working at (Grade F) level.

- [ ] read information from a stem-and-leaf diagram
- [ ] find the mode, median and range from a stem-and-leaf diagram
- [ ] list all the outcomes of two independent events and calculate probabilities from lists or tables
- [ ] calculate the probability of an event not happening when the probability of it happening is known
- [ ] draw a pie chart

You are working a (Grade E) level.

- [ ] draw an ordered stem-and-leaf diagram
- [ ] find the mean of a frequency table of discrete data
- [ ] find the mean from a stem-and-leaf diagram
- [ ] predict the expected number of outcomes of an event
- [ ] draw a line of best fit on a scatter diagram
- [ ] recognise the different types of correlation
- [ ] design a data collection sheet
- [ ] draw a frequency polygon for discrete data

You are working at (Grade D) level.

- [ ] find an estimate of the mean from a grouped table of continuous data
- [ ] draw a frequency diagram for continuous data
- [ ] calculate the relative frequency of an event from experimental data
- [ ] interpret a scatter diagram
- [ ] use a line of best fit to predict values
- [ ] design and criticise questions for questionnaires.

You are working at (Grade C) level.

Basic number

Times tables

- It is essential that you know the **times tables** up to 10 × 10.

- If you take away the 'easy' tables for 1, 2, 5 and 10 and then the 'repeated' tables such as 3 × 4 which is the same as 4 × 3, there are only 15 tables facts left to learn!

1	2	3	4	5	6	7	8	9	10
2	4	6	8	10	12	14	16	18	20
3	6	9	12	15	18	21	24	27	30
4	8	12	16	20	24	28	32	36	40
5	10	15	20	25	30	35	40	45	50
6	12	18	24	30	36	42	48	54	60
7	14	21	28	35	42	49	56	63	70
8	16	24	32	40	48	56	64	72	80
9	18	27	36	45	54	63	72	81	90
10	20	30	40	50	60	70	80	90	100

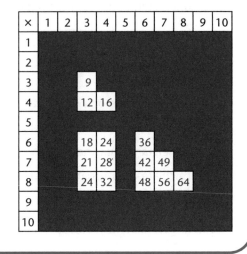

Order of operations and BODMAS

- When you are doing calculations, there is an order of operations that you must follow.

  4 + 6 × 2 is often calculated as 20 because 4 + 6 = 10 and 10 × 2 = 20.

  This is wrong. Multiplication (×) must be done before addition (+), so the correct answer is 4 + 12 = 16.

- The order of operations is given by **BODMAS**, which stands for **B**rackets, **O**rder (or p**O**wer), **D**ivision, **M**ultiplication, **A**ddition, **S**ubtraction.

- Brackets are used to show that parts of the calculation must be done first.

  (4 + 6) × 2 = 20 as the brackets mean 'work out 4 + 6 first'.

Top Tip!

Always use brackets to make calculations clear.

Questions

Grade G

1 a Write down the answer to each part.
  i 4 × 6 **ii** 3 × 7 **iii** 5 × 9
  iv 6 × 8 **v** 9 × 8 **vi** 7 × 7

b Write down the answer to each part.
  i 32 ÷ 4 **ii** 35 ÷ 5 **iii** 63 ÷ 9
  iv 24 ÷ 8 **v** 81 ÷ 9 **vi** 56 ÷ 7

c Write down the answer to each part. There is a remainder in each case.
  i 40 ÷ 7 **ii** 55 ÷ 6 **iii** 43 ÷ 6
  iv 28 ÷ 3 **v** 60 ÷ 9 **vi** 44 ÷ 7

Grade G

2 a Work out each of these.
  i 2 + 3 × 4 **ii** 10 – 2 × 2
  iii 12 + 6 ÷ 2 **iv** 15 – 8 ÷ 4
  v 4 × 8 ÷ 2 **vi** 20 ÷ 5 × 4

b Work out each of these. Remember to work out the brackets first.
  i (5 + 4) × 3 **ii** (9 – 3) × 5
  iii (15 + 7) ÷ 2 **iv** (14 – 2) ÷ 4
  v (4 × 9) ÷ 6 **vi** 20 ÷ (5 × 4)

c Put brackets into these calculations to make them true.
  i 5 × 6 + 1 = 35 **ii** 18 ÷ 2 + 1 = 6
  iii 25 – 10 ÷ 5 = 3 **iv** 20 + 12 ÷ 4 = 8

Remember: You must revise all content from Grade G to the level that you are currently working at.

G

Place value

- The **value** of a digit depends on its position or **place** in the number.

- The **place value** depends on the column heading above the digit.

- The first four column headings are **thousands**, **hundreds**, **tens** and **units**.

  The number 4528 has 4 thousands, 5 hundreds, 2 tens and 8 units.

Thousands	Hundreds	Tens	Units
4	5	2	8

- To make them easier to read, the digits in numbers greater than 9999 are grouped in blocks of three.

  Read 3 456 210 as 'three million, four hundred and fifty-six thousand, two hundred and ten' and 57 643 as 'fifty-seven thousand, six hundred and forty-three'.

G

Rounding

- Most numbers used in everyday life are **rounded**.

  People may say, 'It takes me 30 minutes to drive to work,' or

  'There were forty-two thousand people at the match last weekend.'

- Numbers can be rounded to the **nearest whole number**, the **nearest ten**, the **nearest hundred** and so on.

  76 is rounded to 80 to the nearest 10, 235 is 200 to the nearest hundred and 240 to the nearest ten.

- The convention is that a **halfway** value rounds **upwards**.

  2.5 is rounded to 3 to the nearest whole number.

Top Tip!

Show any 'carried' or 'borrowed' digits clearly.

G-F

Column addition and subtraction

- When adding or subtracting numbers without using a calculator, write the numbers in **columns**.

- Line up the **units digits**.

- **Start** adding or subtracting with the **units digits**.

```
    3 6 3 7          2 ⁷8̶ ¹⁴5̶ ¹4
  +   7 4 8        - 1 3 6 8
    ─────────        ─────────
    4 3 8 5          1 4 8 6
      1   1
```

Questions

Grade G

1 a In the number 3572 what is the value of the digit 5?

b Write the number 27 708 in words.

c Write the number 'Two million, four hundred and six thousand, five hundred and two' in numerals.

Grade G

2 a Round these numbers to the nearest ten.

i 59   **ii** 142   **iii** 45

b Round these numbers to the nearest hundred.

i 682   **ii** 732   **iii** 1250

Grade F

3 Work out the following.

a   2158
     + 3672

b   4215
     − 1637

Remember: You must revise all content from Grade G to the level that you are currently working at.

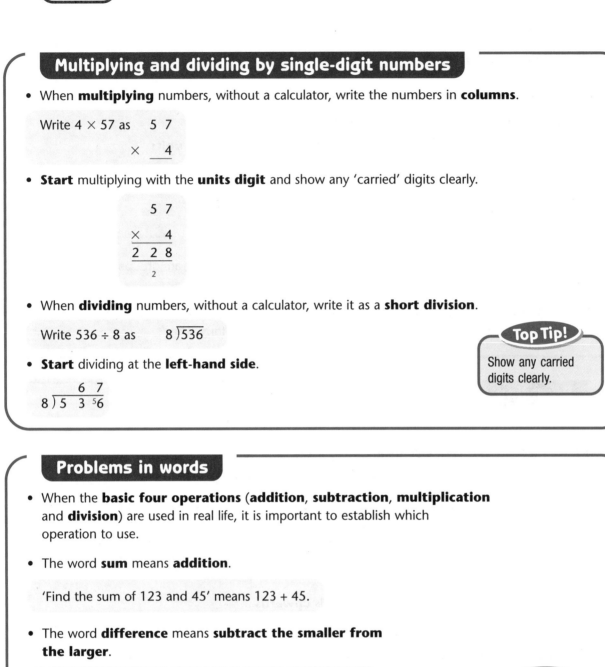

Multiplying and dividing by single-digit numbers

G–F

- When **multiplying** numbers, without a calculator, write the numbers in **columns**.

  Write 4 × 57 as   5 7

                       ×   4

- **Start** multiplying with the **units digit** and show any 'carried' digits clearly.

  ```
      5 7
  ×     4
    2 2 8
      2
  ```

- When **dividing** numbers, without a calculator, write it as a **short division**.

  Write 536 ÷ 8 as   8)536

- **Start** dividing at the **left-hand side**.

  ```
        6 7
  8 ) 5 3 ⁵6
  ```

Top Tip!

Show any carried digits clearly.

Problems in words

G–E

- When the **basic four operations** (**addition**, **subtraction**, **multiplication** and **division**) are used in real life, it is important to establish which operation to use.

- The word **sum** means **addition**.

  'Find the sum of 123 and 45' means 123 + 45.

- The word **difference** means **subtract the smaller from the larger**.

  'Find the difference between 36 and 98' means 98 – 36.

- The word **product** means **multiply**.

  'Find the product of 8 and 26' means 8 × 26.

Top Tip!

Write the necessary calculation in columns.

Questions

Grade F

1 a Work these out.
  i 7 × 39   **ii** 6 × 54

  b Work these out.
  i 208 ÷ 8   **ii** 238 ÷ 7

  c How many days are there in 12 weeks?

  d 180 plants are packed into trays of six plants. How many trays will there be?

Grade F

2 a Find the sum of 382 and 161.

  b Find the difference between 164 and 57.

  c Find the product of 8 and 62.

  d 364 fans travelled to a football match in seven coaches.
  Each coach was full and had the same number of seats.
  How many fans were there in each coach?

Fractions

G

Fractions of a shape

- A **fraction** is part of a whole.

- The number of parts the shape is divided into is called the **denominator** and is the bottom of the fraction.

- The number of parts required is called the **numerator** and is the top of the fraction.

This shape is divided into eight parts and five of them are shaded.

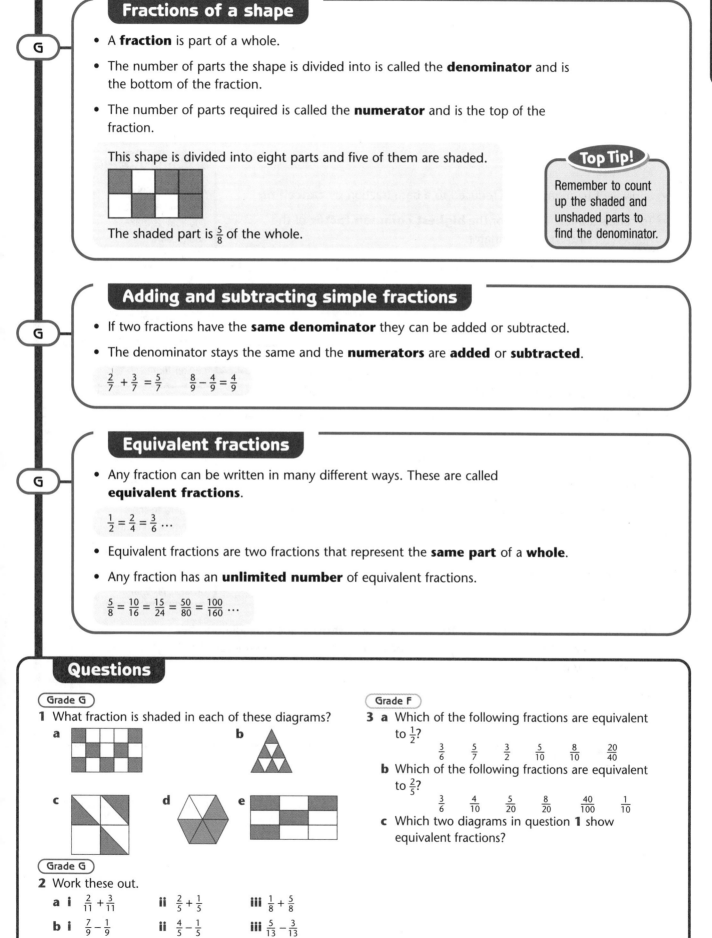

The shaded part is $\frac{5}{8}$ of the whole.

Top Tip!

Remember to count up the shaded and unshaded parts to find the denominator.

G

Adding and subtracting simple fractions

- If two fractions have the **same denominator** they can be added or subtracted.

- The denominator stays the same and the **numerators** are **added** or **subtracted**.

$$\frac{2}{7} + \frac{3}{7} = \frac{5}{7} \qquad \frac{8}{9} - \frac{4}{9} = \frac{4}{9}$$

G

Equivalent fractions

- Any fraction can be written in many different ways. These are called **equivalent fractions**.

$$\frac{1}{2} = \frac{2}{4} = \frac{3}{6} \cdots$$

- Equivalent fractions are two fractions that represent the **same part** of a **whole**.

- Any fraction has an **unlimited number** of equivalent fractions.

$$\frac{5}{8} = \frac{10}{16} = \frac{15}{24} = \frac{50}{80} = \frac{100}{160} \cdots$$

Questions

Grade G

1 What fraction is shaded in each of these diagrams?

a

b

c

d

e

Grade G

2 Work these out.

a i $\frac{2}{11} + \frac{3}{11}$ ii $\frac{2}{5} + \frac{1}{5}$ iii $\frac{1}{8} + \frac{5}{8}$

b i $\frac{7}{9} - \frac{1}{9}$ ii $\frac{4}{5} - \frac{1}{5}$ iii $\frac{5}{13} - \frac{3}{13}$

Grade F

3 a Which of the following fractions are equivalent to $\frac{1}{2}$?

$\frac{3}{6}$ $\frac{5}{7}$ $\frac{3}{2}$ $\frac{5}{10}$ $\frac{8}{10}$ $\frac{20}{40}$

b Which of the following fractions are equivalent to $\frac{2}{5}$?

$\frac{3}{6}$ $\frac{4}{10}$ $\frac{5}{20}$ $\frac{8}{20}$ $\frac{40}{100}$ $\frac{1}{10}$

c Which two diagrams in question **1** show equivalent fractions?

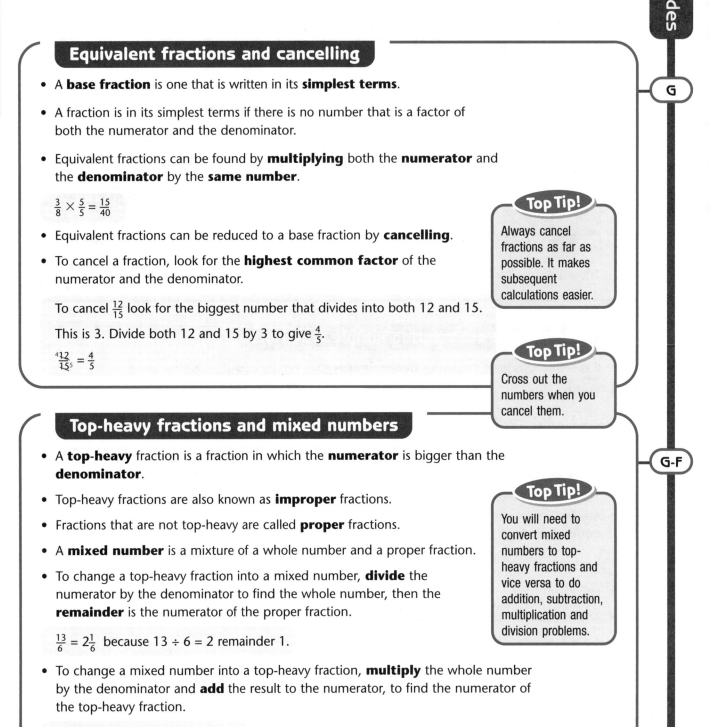

Remember: You must revise all content from Grade G to the level that you are currently working at.

Equivalent fractions and cancelling

G

- A **base fraction** is one that is written in its **simplest terms**.

- A fraction is in its simplest terms if there is no number that is a factor of both the numerator and the denominator.

- Equivalent fractions can be found by **multiplying** both the **numerator** and the **denominator** by the **same number**.

$$\frac{3}{8} \times \frac{5}{5} = \frac{15}{40}$$

- Equivalent fractions can be reduced to a base fraction by **cancelling**.

- To cancel a fraction, look for the **highest common factor** of the numerator and the denominator.

To cancel $\frac{12}{15}$ look for the biggest number that divides into both 12 and 15.

This is 3. Divide both 12 and 15 by 3 to give $\frac{4}{5}$.

$$\frac{{}^4\cancel{12}}{\cancel{15}_5} = \frac{4}{5}$$

> **Top Tip!**
> Always cancel fractions as far as possible. It makes subsequent calculations easier.

> **Top Tip!**
> Cross out the numbers when you cancel them.

Top-heavy fractions and mixed numbers

G-F

- A **top-heavy** fraction is a fraction in which the **numerator** is bigger than the **denominator**.

- Top-heavy fractions are also known as **improper** fractions.

- Fractions that are not top-heavy are called **proper** fractions.

- A **mixed number** is a mixture of a whole number and a proper fraction.

- To change a top-heavy fraction into a mixed number, **divide** the numerator by the denominator to find the whole number, then the **remainder** is the numerator of the proper fraction.

$$\frac{13}{6} = 2\frac{1}{6}$$ because $13 \div 6 = 2$ remainder 1.

- To change a mixed number into a top-heavy fraction, **multiply** the whole number by the denominator and **add** the result to the numerator, to find the numerator of the top-heavy fraction.

$$3\frac{3}{4} = \frac{15}{4}$$ because $4 \times 3 + 3 = 15$.

> **Top Tip!**
> You will need to convert mixed numbers to top-heavy fractions and vice versa to do addition, subtraction, multiplication and division problems.

Questions

Grade G

1 a Find the numbers missing from the boxes.

i $\frac{3}{5} \to \frac{\times 6}{\times 6} = \frac{\square}{30}$ **ii** $\frac{2}{3} \to \frac{\times 8}{\times 8} = \frac{16}{\square}$ **iii** $\frac{5}{8} = \frac{\square}{32}$

b Cancel each of the following fractions.

i $\frac{4}{10}$ **ii** $\frac{3}{15}$ **iii** $\frac{6}{20}$

iv $\frac{15}{25}$ **v** $\frac{15}{21}$

Grade F

2 a Change the following top-heavy fractions into mixed numbers.

i $\frac{12}{5}$ **ii** $\frac{13}{4}$ **iii** $\frac{16}{7}$

iv $\frac{21}{8}$ **v** $\frac{17}{3}$

b Change the following mixed numbers into top-heavy fractions.

i $2\frac{2}{3}$ **ii** $4\frac{3}{5}$ **iii** $2\frac{6}{7}$

iv $2\frac{1}{4}$ **v** $3\frac{5}{8}$

 **Remember:** You must revise all content from Grade G to the level that you are currently working at.

Adding and subtracting fractions

- When two fractions are added (or subtracted) there are four possible outcomes.
  - The answer will be a proper fraction that does not need to be cancelled.

$$\frac{1}{7} + \frac{3}{7} = \frac{4}{7}$$

  - The answer will be a proper fraction that needs to be cancelled.

$$\frac{5}{9} - \frac{2}{9} = \frac{3}{9} = \frac{1}{3}$$

  - The answer will be a top-heavy fraction that does not need to be cancelled, so the fraction is converted to a mixed number.

$$\frac{3}{5} + \frac{4}{5} = \frac{7}{5} = 1\frac{2}{5}$$

  - The answer will be a top-heavy fraction that needs to be cancelled, then the fraction is converted to a mixed number.

$$\frac{7}{9} + \frac{5}{9} = \frac{12}{9} = \frac{4}{3} = 1\frac{1}{3}$$

> **Top Tip!**
> Always cancel before converting the top-heavy fraction into a mixed number.

Finding a fraction of a quantity

- To find a **fraction of a quantity** just **multiply** the fraction by the quantity.

To find $\frac{3}{5}$ of 25, $\frac{1}{5}$ of 25 = 5, so $\frac{3}{5} \times 25 = 3 \times 5 = 15$.

Find $\frac{2}{7}$ of 49 kg. $\frac{1}{7}$ of 49 = 7
                               $\frac{2}{7}$ of 49 = 2 × 7 = 14

> **Top Tip!**
> Divide the quantity by the denominator to find the unit fraction of the quantity, then multiply this unit fraction by the numerator.

- To compare fractions of numbers of quantities, work out the fractions first.

Which is the larger number, $\frac{2}{5}$ of 40 or $\frac{3}{7}$ of 35?
$\frac{1}{5}$ of 40 = 8
$\frac{2}{5}$ of 40 = 2 × 8 = 16
$\frac{1}{7}$ of 35 = 5
$\frac{3}{7}$ of 35 = 3 × 5 = 15   So $\frac{2}{5}$ of 40 is larger.

Questions

Grade G

1 Add or subtract the following fractions. Cancel the answers and/or make into mixed numbers if necessary.

a i $\frac{1}{5} + \frac{3}{5}$ **ii** $\frac{5}{7} - \frac{3}{7}$ **iii** $\frac{1}{9} + \frac{4}{9}$
  iv $\frac{2}{3} - \frac{1}{3}$

b i $\frac{1}{6} + \frac{1}{6}$ **ii** $\frac{3}{10} - \frac{1}{10}$ **iii** $\frac{1}{9} + \frac{5}{9}$
  iv $\frac{5}{12} - \frac{1}{12}$

c i $\frac{6}{9} + \frac{7}{9}$ **ii** $\frac{5}{13} + \frac{10}{13}$ **iii** $\frac{7}{11} + \frac{8}{11}$
  iv $\frac{2}{3} + \frac{2}{3}$

d i $\frac{7}{8} + \frac{5}{8}$ **ii** $\frac{7}{10} + \frac{9}{10}$ **iii** $\frac{5}{9} + \frac{7}{9}$
  iv $\frac{11}{12} + \frac{7}{12}$

Grade F

2 Calculate the following.

  a $\frac{2}{5}$ of 20
  b $\frac{5}{8}$ of 32
  c $\frac{5}{6}$ of 30
  d $\frac{3}{4}$ of £300
  e $\frac{2}{9}$ of 81 kg
  f $\frac{2}{3}$ of 6 hours
  g Which is the larger number, $\frac{2}{3}$ of 21 or $\frac{3}{4}$ of 20?
  h Which is the larger number, $\frac{4}{7}$ of 63 or $\frac{7}{8}$ of 40?

Remember: You must revise all content from Grade G to the level that you are currently working at.

Multiplying fractions

- To multiply two fractions, simply multiply the numerators to get the new numerator and multiply the denominators to get the new denominator.

$$\frac{3}{5} \times \frac{2}{7} = \frac{3 \times 2}{5 \times 7} = \frac{6}{35}$$

- If possible **cancel** numbers on the top and bottom before multiplying.

In $\frac{5}{6} \times \frac{9}{10}$ cancel 5 from 5 and 10 and cancel 3 from 6 and 9.

$$\frac{{}^{1}5}{{}_{2}6} \times \frac{9^{3}}{10^{2}} = \frac{3}{4}$$

Top Tip!
Always cancel before multiplying, as it makes the calculations easier and you will not have to cancel the final answer.

— E

One quantity as a fraction of another

- To write one quantity as a fraction of another, write the **first quantity** as the **numerator** and the **second quantity** as the **denominator**.

What is £8 as a fraction of £20?

Write as $\frac{8}{20}$ then cancel to $\frac{2}{5}$.

Top Tip!
Examination questions often ask for the fraction to be given in its simplest form. This means it has to be cancelled down.

— D

Problems in words

- In examinations, most fraction problems will be set out as real-life problems expressed in words.

- Decide on the calculation, write it down then work it out.

John eats a quarter of a cake and Mary eats half of what is left. What fraction of the original cake is left?

John eats $\frac{1}{4}$. Mary eats $\frac{1}{2} \times \frac{3}{4} = \frac{3}{8}$.

So $\frac{3}{8}$ of the cake is left.

— F-D

Questions

Grade E

1 Multiply the following fractions. Give the answers in their simplest form.

a $\frac{1}{3} \times \frac{3}{5}$

b $\frac{5}{8} \times \frac{3}{5}$

c $\frac{1}{2} \times \frac{4}{9}$

d $\frac{2}{3} \times \frac{9}{10}$

e $\frac{2}{9} \times \frac{1}{8}$

f $\frac{3}{8} \times \frac{4}{15}$

g $\frac{4}{9} \times \frac{3}{8}$

h $\frac{7}{8} \times \frac{10}{21}$

Grade D

2 a What fraction of 25 is 10?

b In a class of 28 students, 21 are right-handed. What fraction is this?

c What fraction is 20 minutes of an hour?

Grade D

3 a In a mixed box of tapes and CDs, $\frac{2}{3}$ of the items were CDs. What fraction were tapes?

b In a packet of toffees $\frac{1}{2}$ are plain toffees, $\frac{3}{8}$ are nut toffees and the rest are treacle toffees. What fraction are treacle toffees?

c Ken earns £400 a week. One week he earns a bonus of $\frac{1}{5}$ of his wages. How much does he earn that week?

Rational numbers and reciprocals

Rational numbers

- A **rational number** is any number that can be expressed as a **fraction**.

- Some fractions result in **terminating decimals**.

  $\frac{1}{16} = 0.0625$, $\frac{3}{64} = 0.046\,875$, $\frac{7}{40} = 0.175$,

  $\frac{19}{1000} = 0.019$, so all give terminating decimals.

- Some fractions result in **recurring** decimals.

  $\frac{1}{3} = 0.3333...$

- **Recurrence** is shown by dots over the recurring digit or digits.

  0.3333... becomes $0.\dot{3}$ in recurrence or dot notation.

  0.277 777... becomes $0.2\dot{7}$ 0.518 518 518... becomes $0.\dot{5}1\dot{8}$

- To convert a fraction into a decimal, just divide the numerator by the denominator.

  $\frac{7}{20} = 7 \div 20 = 0.35$, $\frac{5}{11} = 5 \div 11 = 0.454\,545... = 0.\dot{4}\dot{5}$

Top Tip!

The only fractions that give terminating decimals are those with a denominator that is a power of 2, 5 or 10, or a combination of these.

Converting terminating decimals into fractions

- Depending on the number of places in the decimal, the denominator will be 10, 100, 1000, ...

  $0.7 = \frac{7}{10}$, $0.036 = \frac{36}{1000} = \frac{9}{250}$, $2.56 = \frac{256}{100} = \frac{64}{25} = 2\frac{14}{25}$

Top Tip!

Count the number of decimal places. This is the same as the number of zeros in the denominator.

Finding reciprocals

- The **reciprocal** of a number is the result of **dividing** the number **into 1**.

  The reciprocal of 5 is $\frac{1}{5}$ or $1 \div 5 = 0.2$.

- The reciprocal of a **fraction** is simply the fraction turned **upside down**.

  The reciprocal of $\frac{4}{7}$ is $\frac{7}{4} = 1\frac{3}{4}$.

Questions

Grade C

1 Use recurrence or dot notation to write each of these.

  a 0.363 636...   **b** 0.615 615 615...

  c 0.3666...   **d** $\frac{2}{3}$

  e $\frac{1}{6}$ **f** $\frac{7}{9}$

Grade D

2 Convert the following terminating decimals into fractions. Cancel your answers if possible.

  a 0.8   **b** 0.65   **c** 0.125

  d 2.45   **e** 0.025   **f** 0.888

Grade C

3 a Work out the reciprocal of each number. Give your answers as a terminating or recurring decimals.

    i 4   **ii** 20   **iii** 9

  b Write down the reciprocal of each fraction. Give your answers as mixed numbers.

    i $\frac{7}{8}$ **ii** $\frac{5}{9}$ **iii** $\frac{3}{13}$

Negative numbers

Negative numbers

- The **natural numbers** are 1, 2, 3, 4, 5, ...

- The **counting numbers are** 0, 1, 2, 3, 4, 5, ...

- **Negative numbers** are those numbers below zero such as –1, –2, –8.3.

- Negative numbers are used in everyday life to give **temperatures**, **distances below** certain points (such as sea level) and to indicate **overdrawn bank balances**, for example.

- Negative numbers are also sometimes known as **directed numbers** when they are used with a number line.

Top Tip!

The larger the actual number in a negative number, the smaller it is. i.e. –9 is smaller than –3.

G

The number line

- Negative numbers can be represented on a **number line**.

- Number lines can be **horizontal** or **vertical**.

```
 –7  –6  –5  –4  –3  –2  –1   0   1   2   3   4   5   6   7
             negative                    positive
```

- Negative numbers are to the **left** or **below** zero.

- Positive numbers are to the **right** or **above** zero.

- The further to the **left** or the further **down** the number line, the **smaller** the number.

```
7
6
5
4
3
2
1
0
–1
–2
–3
–4
–5
–6
–7
```

G

Questions

Grade G

1 Complete the following sentences.

a If +£7 means a profit of seven pounds, then ... means a loss of seven pounds.

b If +300 m means 300 metres above sea level, then ... means 40 metres below sea level.

c If –6 means 6 hours before midnight, then 8am the next day would be represented by ... .

d If 40°F is 8 degrees above freezing in the Fahrenheit scale, then 8 degrees below freezing is given by ... °F.

e If –100 km means 100 kilometres south of Leeds, then 50 km north of Leeds would be represented by ... .

f If –100 km means 100 kilometres west of Leeds, then 2 km east of Leeds would be represented by ... .

Grade G

2 Use the number line above to answer this question.

a Put < or > in each statement to make it true.

  i –3 ... 3   **ii** –8 ... –10   **iii** –6 ... –2

b Write down the next two terms of this sequence.

  11, 8, 5, 2, ... , ...

c For each pair of numbers, write down the number that is halfway between them.

  i 2 and 8   **ii** –8 and 2   **iii** –6 and –1

Remember: You must revise all content from Grade G to the level that you are currently working at.

F

Addition and subtraction with negative numbers

- Adding a positive number is a move to the right.

  $-2 + +7 = 5$

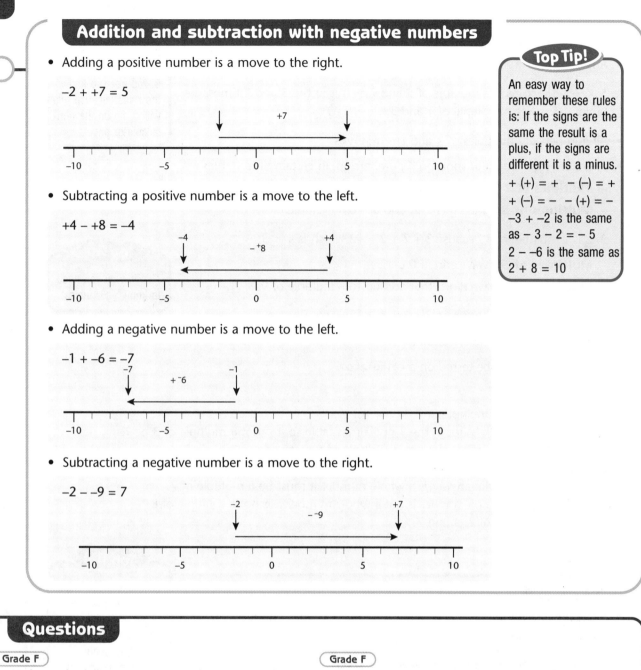

Top Tip!

An easy way to remember these rules is: If the signs are the same the result is a plus, if the signs are different it is a minus.

$+ (+) = +$ $- (-) = +$

$+ (-) = -$ $- (+) = -$

$-3 + -2$ is the same as $-3 - 2 = -5$

$2 - -6$ is the same as $2 + 8 = 10$

- Subtracting a positive number is a move to the left.

  $+4 - +8 = -4$

- Adding a negative number is a move to the left.

  $-1 + -6 = -7$

- Subtracting a negative number is a move to the right.

  $-2 - -9 = 7$

Questions

Grade F

1 Work out the following. Use a number line to help if necessary.

a $6 - 10$

b $-3 - 9$

c $-6 + 8$

d $+5 - 8$

e $-9 + 9$

f $6 - 8 - 2$

g $3 - +9$

h $-2 - -7$

i $-3 - +8 - +3$

j $-9 - -5$

k $7 - -5$

l $-7 + 8 - 9$

m $4 - 6 + 9$

n $-5 - 8 - 9$

o $-6 + -8 + +6$

Grade F

2 a What is the sum of +3, −5 and −8?

b What is the sum of −6, −7 and −11?

c What is the difference between −10 and +11?

d What is the difference between −8 and −14?

Grade F

3 Write down the calculation shown on each number line.

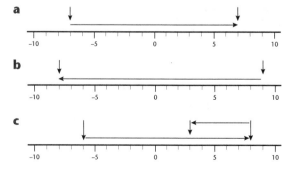

a

b

c

More about number

Multiples

- A **multiple** of a number is any number in the **times table**.

$5 \times 7 = 35$, so 35 is a multiple of 5 and 7 as it is in the 5 and 7 times tables. 35 is also a multiple of **1** and 35.

- All numbers are multiples of **1** and **themselves**.
- There are some rules for spotting if numbers are in certain times tables.
  - **Multiples of 2** (which are also even numbers) always end in 0, 2, 4, 6 or 8.
  - **Multiples of 3** have digits that add up to a multiple of 3.

372 is a multiple of 3 because $3 + 7 + 2 = 12$ which is 4×3.

  - **Multiples of 5** always end in 0 or 5.
  - **Multiples of 9** have digits that add up to a multiple of 9.

738 is a multiple of 9 because $7 + 3 + 8 = 18 = 2 \times 9$.

  - **Multiples of 10** always end in 0.
- Other multiples can be tested on a **calculator**.

To test 196:
As it ends in 6 it is a multiple of 2.
$1 + 9 + 6 = 16$, which is not in the 3 or 9 times table so it is not a multiple of 3 or 9.
It does not end in 0 or 5 so it is not a multiple of 5 or 10.
$196 \div 4 = 49$, so as the answer is a whole number it must be a multiple of 4.
$196 \div 7 = 28$, so as the answer is a whole number it must be a multiple of 7.
$196 \div 8 = 24.5$, so as the answer is not a whole number it is not a multiple of 8.
So 196 is a multiple of 2, 4 and 7.

Top Tip! You might be asked to say, for example, if 729 is a multiple of 9. You need to know the rules because this will come on the non-calculator paper.

F

Top Tip! This is why you need to know your tables!

Factors

- A **factor** of a number is any whole number that divides into it exactly.
The factors of 20 are {1, 2, 4, 5, 10, 20}.
- **1** is always a factor and so is the **number itself**.
- When you find one factor there is always **another** that goes with it, unless you are investigating a square number.

$1 \times 12 = 12$, $2 \times 6 = 12$, $3 \times 4 = 12$ so the factors of 12 are {1, 2, 3, 4, 6, 12}.
$1 \times 16 = 16$, $2 \times 8 = 16$, $4 \times 4 = 16$, so the factors of 16 are {1, 2, 4, 8, 16}.

Top Tip! On the non-calculator paper, you may be asked to find the factors of numbers up to 60. Any bigger numbers will be on the calculator paper.

G-F

Questions

Grade F

1 a Write down the first five multiples of each number.
 i 6 **ii** 13 **iii** 25
b From the list at the top of the next column, identify those numbers that are multiples of:
 i 2 **ii** 3 **iii** 5
 iv 9 **v** 10.

| 121 | 250 | 62 | 78 | 90 | 85 | 108 |
| 144 | 35 | 81 | 96 | 120 | 333 | 125 |

c Is 819 a multiple of 9?

Grade F

2 a Write down the factors of each number.
 i 24 **ii** 15 **iii** 50 **iv** 40
b Use a calculator to find the factors of 144.

Primes and squares

Prime numbers

- A **prime number** is a number with only two **factors**.

  17 has only the factors 1 and 17, 3 has only the factors 1 and 3.

- The factors of a prime number are always **1** and the number **itself**.

- There is no rule or pattern for spotting prime numbers, you will have to **learn them**.

  The prime numbers up to 50 are:

  2, 3, 5, 7, 11, 13, 17, 19, 23, 29, 31, 37, 41, 43 and 47.

- All prime numbers are **odd**, **except for 2**, which is the **only even** prime number.

> **Top Tip!**
> 1 is *not* a prime number as it only has one factor (1).

Square numbers

- The sequence 1, 4, 9, 16, 25, 36, ... is called the **sequence of square numbers**.

- The square numbers can be represented by **square patterns**.

1×1 2×2 3×3 4×4 5×5

- Square numbers can be calculated as 1×1, 2×2, 3×3 and so on.

- 1×1, 2×2, 3×3 ... can be written as numbers to the power 2: 1^2, 2^2, 3^2, ...

- The power 2 is referred to as the number **squared**.

  Read 7^2 as '7 squared'.

> **Top Tip!**
> You are expected to know the square numbers up to 15×15 (= 225).

Questions

Grade E

1 a Write down the factors of each number.

   i 18   **ii** 19   **iii** 20

  b Which of the numbers in part (**a**) is a prime number?

  c From the list below, identify those that are prime numbers.

     21   25   61   79   9
     83   17   41   35   81
     29   1   33   15

  d What is the only even prime number?

Grade E

2 a Continue the following sequence to 10 terms.

    1, 4, 9, 16, 25, ... , ... , ... , ... , ...

  b Work out the value of each number.

   i 11^2 **ii** 12^2 **iii** 13^2

   iv 14^2 **v** 15^2

  c Continue the following number pattern for three more lines.

$$1 + 3 = 4$$
$$1 + 3 + 5 = 9$$
$$1 + 3 + 5 + 7 = 16$$

Roots and powers

Square roots

- The **square root** of a given number is a number that, when multiplied by itself, produces the given number.

  The square root of 16 is 4 as $4 \times 4 = 16$.

- Square roots can also be **negative**.

  -4×-4 also equals 16.

- A square root is represented by the **symbol** $\sqrt{}$.

  $\sqrt{25} = 5$

- Calculators have a '**square root**' **button**.

- Taking a square root is the **inverse operation** to squaring.

F

Powers

- **Powers** are a convenient way of writing **repeated multiplications**.

  $3 \times 3 \times 3 \times 3 \times 3 \times 3 = 3^6$, which is read as '3 to the power 6'.

- Powers are also called **indices** (singular **index**).

- Most calculators have **power buttons**.

  x^2 x^3 x^y

- The **power 2** has a special name: '**squared**'.

- The **power 3** has a special name: '**cubed**'.

- The **inverse operation** of cubing is taking the **cube root**, which is represented by the symbol $\sqrt[3]{}$.

  $\sqrt[3]{8} = 2$, $\sqrt[3]{27} = 3$

F-C

Questions

Grade F

1 a Write down the positive square root of each of these numbers.

   i 81   **ii** 64   **iii** 25

b For each statement, write down two values of x that make it true.

   i $x^2 = 4$ **ii** $x^2 = 16$ **iii** $x^2 = 100$

c Use a calculator to work out these square roots.

   i $\sqrt{576}$ **ii** $\sqrt{6.25}$ **iii** $\sqrt{37.21}$

Grade D

2 a Write down the values of these numbers.

   i 3^3 **ii** 4^3 **iii** 10^3

b Write these numbers, using power notation.

   i $4 \times 4 \times 4 \times 4 \times 4$

   ii $6 \times 6 \times 6 \times 6 \times 6 \times 6$

   iii $10 \times 10 \times 10 \times 10$

   iv $2 \times 2 \times 2 \times 2 \times 2 \times 2 \times 2$

c Use a calculator to work out the values of the powers in part (**b**).

d Continue the sequence of the powers of 2 up to 10 terms.

   2, 4, 8, 16, 32, ... , ... , ... , ... , ...

Powers of 10

Powers of 10

- The **decimal** system, which is the basis of our number system, is based on **powers of 10**.

- **Column headings** such as hundreds, tens and units are **powers of 10**.

The number 346.32 could be written as:

100	10	1	$\frac{1}{10}$	$\frac{1}{100}$
3	4	6 •	3	2

or

10^2	10^1	10^0	10^{-1}	10^{-2}
3	4	6 •	3	2

> **Top Tip!**
> The positive power of 10 and the number of zeros after the number are the same, for example, $10^5 = 100\,000$, and 1 million is 10^6.

- The decimal point separates the whole numbers from the decimal fractions.

- The column headings after the decimal point are **tenths**, **hundredths** or 10^{-1}, 10^{-2}.

Multiplying and dividing by powers of 10

- When you **multiply** by a power of 10, the digits of the number move to the **left**.
- When you **divide** by a power of 10, the digits move to the **right**.

> **Top Tip!**
> Although, strictly speaking, the digits move, in reality it is easier to think of it as moving the decimal point.

$34.5 \times 10 = 345$

100	10	1	$\frac{1}{10}$
	3	4 •	5
3	4	5 •	0

$\times 10$

$45.9 \div 10^2 = 0.459$

10	1		$\frac{1}{10}$	$\frac{1}{100}$	$\frac{1}{1000}$
4	5	•	9		
		•	4	5	

$\div 10^2$

- The number of **places the digits move** depend on the number of **zeros** or the power of 10.

Multiplying and dividing multiples of powers of 10

- When you **multiply** together two multiples of powers of 10, just multiply the non-zero digits and write the **total** of the zeros in both numbers at the end.

$200 \times 4000 = 800\,000$, $500 \times 60 = 30\,000$

- When you **divide** one multiple of powers of 10 by another, just divide the non-zero digits and write the **difference** in the zeros in both numbers at the end.

$8000 \div 20 = 400$, $20\,000 \div 40 = 500$

Questions

Grade E

1 Write down the answers.

  a 8×100 **b** 6.4×10

  c 2.5×10^2 **d** 0.3×10^3

  e 7.6×100 **f** 3.25×1000

  g $64 \div 10$ **h** $2.8 \div 100$

  i $390 \div 10^2$ **j** $0.75 \div 10$

  k $34 \div 10^3$ **l** $9.4 \div 10$

Grade E

2 Write down the answers.

  a 3000×200 **b** 40×300

  c 2000×70 **d** $4000 \div 20$

  e $6000 \div 300$ **f** $40\,000 \div 800$

Prime factors

Prime factors

- When a number is written as a product of **prime factors** it is written as a multiplication consisting only of prime numbers.

  $30 = 2 \times 3 \times 5$, $50 = 2 \times 5 \times 5$ or 2×5^2

- There are two ways to find prime factors: the **division method** and the **tree method**.

The division method to find the prime factors of 24

Divide by prime numbers until the answer is a prime number.

```
2 ) 2  4
2 ) 1  2
2 )    6
       3
```

So $24 = 2 \times 2 \times 2 \times 3$

> **Top Tip!**
>
> If a number is even then 2 is an obvious choice as a divisor or part of the product, then look at 3, 5, 7, ...

The tree method to find the prime factors of 76

Keep splitting numbers into products until there are prime numbers at the ends of all the branches.

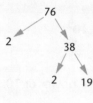

So $76 = 2 \times 2 \times 19$

- Products of prime factors can be expressed in **index form**.

  $24 = 2 \times 2 \times 2 \times 3 = 2^3 \times 3$

  $50 = 2 \times 5 \times 5 = 2 \times 5^2$

  $76 = 2 \times 2 \times 19 = 2^2 \times 19$

Questions

Grade C

1 What numbers are represented by these products of prime factors?

   a $2 \times 3 \times 5$ **b** $2 \times 2 \times 3 \times 7$

   c $2 \times 5 \times 13$ **d** $2^2 \times 3^2$

   e $2^3 \times 5$ **f** $2^2 \times 3 \times 5^2$

Grade C

2 Use the division method to find the product of prime factors for each of these numbers. Give your answers in index form if possible.

   a 20   **b** 45   **c** 64   **d** 120

Grade C

3 Use the tree method to find the product of prime factors for each of these numbers. Give your answers in index form if possible.

   a 16   **b** 42   **c** 70   **d** 200

LCM and HCF

C

Lowest common multiple

- The **lowest common multiple** (**LCM**) of two numbers is the smallest number that appears in the times tables of both of the two numbers.

  The LCM of 6 and 7 is 42, the LCM of 8 and 20 is 40.

- There are two ways to find the LCM: the **prime factor method** and the **list method**.

The prime factor method

Find the LCM of 24 and 40.

Write 24 and 40 as products of prime factors: $24 = 2^3 \times 3$, $40 = 2^3 \times 5$

Now find the smallest product of prime factors that includes all the prime factors of 24 and 40.

This is $2^3 \times 3 \times 5 = 120$.

So the LCM of 24 and 40 is 120.

The list method

Find the LCM of 16 and 20.

Write out the 16 and 20 times tables, continuing until there is number that appears in both lists (a common multiple).

16 times table: 16, 32, 48, 64, ⑧⓪, 96, 112, 128, ...

20 times table: 20, 40, 60, ⑧⓪, 100, 120, ...

So the LCM of 16 and 20 is 80.

Highest common factor

C

- The **highest common factor** (**HCF**) of two numbers is the **biggest** number that **divides exactly** into the two numbers.

  The HCF of 16 and 20 is 4, the HCF of 18 and 42 is 6.

- There are two ways to find the HCF: the **prime factor method** and the **list method**.

The prime factor method

Find the HCF of 45 and 108.

Write 45 and 108 as products of prime factors: $45 = 3^2 \times 5$, $108 = 2^2 \times 3^3$

Now find the biggest product of prime factors that is included in the prime factors of 45 and 108.

This is $3^2 = 9$.

So the HCF of 45 and 108 is 9.

The list method

Find the HCF of 36 and 90.

Write out the factors of 36 and 90 then pick out the biggest number that appears in both lists.

Factors of 36: {1, 2, 3, 4, 6, 9, 12, ⑱, 36}

Factors of 90: {1, 2, 3, 5, 6, 9, 10, 15, ⑱, 30, 45, 90}

So the HCF of 36 and 90 is 18.

Questions

Grade C

1 a Find the LCM of each pair of numbers.

    i 5 and 6   **ii** 3 and 7   **iii** 3 and 13

b Describe a connection between the LCM and the original numbers in part (**a**).

c Find the LCM of each pair of numbers.

    i 6 and 9   **ii** 8 and 20   **iii** 15 and 25

Grade C

2 Find the HCF of the numbers in each pair.

    a 12 and 30   **b** 18 and 40   **c** 15 and 50

    d 16 and 80   **e** 24 and 60   **f** 12 and 25

Grade C

3 Find the LCM of the numbers in each pair.

    a 12 and 30   **b** 15 and 50   **c** 24 and 60

Powers

Multiplying powers

- When two powers of the **same base number** are **multiplied** together, then the new power is the **sum** of the original powers.

  $4 \times 8 = 32$, but $4 \times 8 = 2^2 \times 2^3 = 2^5 = 32$

C

Dividing powers

- When two powers of the **same base number** are **divided**, then the new power is the **difference** of the original powers.

  $243 \div 9 = 27$, but $243 \div 9 = 3^5 \div 3^2 = 3^3 = 27$

- Any number to the power 1 is just the number itself.

  $32 \div 16 = 2$, but $32 \div 16 = 2^5 \div 2^4 = 2^{5-4} = 2^1$

- Any number to the power 0 is always 1.

  $27 \div 27 = 1$, but $27 \div 27 = 3^3 \div 3^3 = 3^{3-3} = 3^0$

Top Tip!

You are often asked to write an expression such as $5^2 \times 5^3$ as a single power of 5, which would be 5^5. You do not have to work out the actual value unless you are asked to do so, and this would not be on the non-calculator paper anyway.

C

Multiplying and dividing powers with letters

- When you are working with **numbers**, using the **rules for multiplying and dividing powers** helps to simplify calculations but they can always be evaluated as numerical answers.

- When you use algebraic unknowns, or letters, as the base the **expression cannot be simplified**.

- Algebraic powers will not be tested in module 3 but they follow exactly the same rules as above.

  $x^2 \times x^6 = x^8$ $x^9 \div x^6 = x^3$

Top Tip!

Don't get confused with ordinary numbers when powers are involved. The rules above only apply to powers, so
$2x^3 \times 3x^2$
$= 2 \times 3 \times x^3 \times x^2$
$= 6x^5$ *not* $5x^5$.

C

Questions

Grade C

1 a Write each of the following as a single power of 2.

  i $2^3 \times 2^4$ **ii** $2^4 \times 2^5$ **iii** $2^3 \times 2^3$

b Write each of the following as a single power of 3.

  i $3^5 \div 3^2$ **ii** $3^8 \div 3^4$ **iii** $3^6 \div 3^3$

c Write each of the following as a single power of x.

  i $x^6 \times x^3$ **ii** $x^5 \times x^4$ **iii** $x^6 \times x^5$

d Write each of the following as a single power of x.

  i $x^7 \div x^3$ **ii** $x^9 \div x^3$ **iii** $x^5 \div x^2$

e Which of the following is the algebraic rule for $x^n \times x^m$?

  i x^{nm} **ii** x^{n+m} **iii** $(m+n)x$

f Which of the following is the algebraic rule for $x^n \div x^m$?

  i x^{n-m} **ii** $x^{n \div m}$ **iii** $(m-n)x$

Grade C

2 Write down the value of each number.

  a 4^0 **b** 7^1

  c $8^5 \div 8^5$ **d** $6^4 \div 6^3$

Grade C

3 a Which of the following is the correct answer to $2x^2 \times 5x^5$?

  i $7x^{10}$ **ii** $10x^7$ **iii** $7x^7$ **iv** $10x^7$

b Which of the following is the correct answer to $12x^6 \div 3x^2$?

  i $9x^3$ **ii** $4x^3$ **iii** $4x^4$ **iv** $9x^4$

Number skills

Long multiplication

- **Long multiplication** is used to multiply two numbers that both have more than one digit.

- There are several methods for long multiplication. The **four most common** are:
  - **Napier's bones** or **Chinese multiplication**
  - the **column** method
  - the **expanded column** method
  - the **box** method.

> **Top Tip!**
> In the GCSE non-calculator paper, you are only expected to multiply a two-digit number by a three-digit number at the most. Any calculations involving bigger numbers will be on the calculator paper.

Napier's bones or Chinese multiplication

Work out 27 × 42.

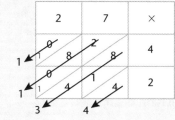

So 27 × 42 = 1134.
Note the carried digits.

The column method

Work out 36 × 28.

```
        3 6
   ×    2 8
      2 8 8
        4
      7 2 0
      1
    1 0 0 8
    1 1
```

So 36 × 48 = 1008.

> **Top Tip!**
> This is the traditional method but it can be confusing, with all the carried digits and the need to remember to put down a zero.

Expanded column method

Work out 47 × 52.

```
      5 2            5 2          2 0 8 0
  ×   4 0        ×     7      +     3 6 4
  2 0 8 0        3 6 4          2 4 4 4
                     1                1
```

So 47 × 52 = 2444.

> **Top Tip!**
> This method makes the calculation into two short multiplications.

The box method

Work out 23 × 45.

×	20	3
40	800	120
5	100	15

```
    8 0 0
    1 2 0
    1 0 0
  +   1 5
  1 0 3 5
      1
```

So 23 × 45 = 1035.

> **Top Tip!**
> Decide which method you prefer and stick to it. It doesn't matter which method you use in the GCSE as long as you get the answer right.

Questions

Grade F

1 Use whichever method you prefer to work these out.

a 24 × 23 **b** 61 × 52 **c** 63 × 31
d 78 × 34 **e** 147 × 43 **f** 265 × 26

Remember: You must revise all content from Grade G to the level that you are currently working at.

Long division

- Long division is used to divide a number with three or more digits by a number with two or more digits.

- There are two methods for long division:
    - the **traditional** or **Italian** method
    - the **repeated subtraction** or **chunking** method.

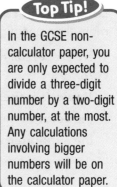

Top Tip!

In the GCSE non-calculator paper, you are only expected to divide a three-digit number by a two-digit number, at the most. Any calculations involving bigger numbers will be on the calculator paper.

The traditional or Italian method

Work out 864 ÷ 24.

```
        3 6
  24 ) 8 6 4
      7 2
      ‾‾‾‾‾
      1 4 4
      1 4 4
      ‾‾‾‾‾
          0
```

1 How many 24s in 86?
  There are 3 and 3 × 14 = 72.

2 86 − 72 = 14 bring the 4 down to make 144.

3 How many 24s in 144? There are 6 exactly.

This is basically a way of showing a 'short' division neatly.

```
        3 6
  24 ) 8 ⁸6¹⁴4
```
So 864 ÷ 24 = 36.

Repeated subtraction or chunking

Work out 1564 ÷ 34.

1 Start by writing down the easier multiples of 34.

  1 × 34 = 34, 2 × 34 = 68, 10 × 34 = 340, 20 × 34 = 680

2 Keep on subtracting multiples of 34, taking away the biggest you can each time.

```
  1 5 6 4
    6 8 0      20 ×
  ‾‾‾‾‾‾‾
    8 8 4
    6 8 0      20 ×
  ‾‾‾‾‾‾‾
    2 0 4
      6 8      2 ×
    ‾‾‾‾‾
    1 3 6
      6 8      2 ×
    ‾‾‾‾‾
      6 8
      6 8      2 ×
    ‾‾‾‾‾
        0      46 ×
```

Top Tip!

Decide which method you prefer and stick to it. It doesn't matter which method you use in the GCSE as long as you get the answer right.

3 When the answer is 0 or less than 34, add up how many have been taken away.
  So 1564 ÷ 34 = 46.

Questions

(Grade F)

1 Use whichever method you prefer to work these out.

a 572 ÷ 22   **b** 528 ÷ 33   **c** 988 ÷ 26
d 704 ÷ 16   **e** 672 ÷ 24   **f** 637 ÷ 13

Remember: You must revise all content from Grade G to the level that you are currently working at.

F

Real-life problems

- In a GCSE exam, long multiplication and long division are usually given in the context of **real-life problems**.

- You will need to **identify** the problem as a **multiplication** or **division** and then work it out by your preferred method.

- In the GCSE non-calculator paper, questions often ask such things as 'How many coaches are needed?' and the calculation gives a remainder. Remember that an extra coach will be needed to carry the remaining passengers. You cannot have a 'bit of a coach'.

> **Top Tip!**
> Show your working clearly because even if you make a small arithmetical error you will still get marks for method.

A café uses 950 eggs per day. The eggs come in trays of 24. How many trays of eggs will the café need?

This is a division problem. Use the traditional method.

```
        3 9
24 ) 9  5  0
     7  2
     ‾‾‾‾‾‾
     2  3  0
     2  1  6
     ‾‾‾‾‾‾
        1  4
```

The answer to the calculation 950 ÷ 24 is 39 remainder 14.

The café will need 40 trays and will have 10 eggs left over.

F

Decimal places

- The number of decimal places in a number is just the **number of digits after the decimal point**.

  2.34 has two decimal places (2 dp), 3.068 has three decimal places (3 dp).

- You need to be able to round numbers to one, two or three decimal places (1 dp, 2 dp, 3 dp).

- To round a number to one decimal place, look at the digit in the **second** decimal place. If it is less than 5 remove the unwanted digits. If it is 5 or more add 1 on to the digit in the **first** decimal place.

- Use the same method to round to two or three decimal places.

  3.5629 is 3.6 to 1 dp, 3.56 to 2 dp and 3.563 to 3 dp.

> **Top Tip!**
> There will always be at least one question in your GCSE that asks you to round a number.

Questions

Grade F

1 a There are 945 students in a school. There are 27 students in each tutor group. How many tutor groups are there?

b To raise money for charity one tutor group decides that all 27 members will donate the cost of a KitKat. If a KitKat costs 42p, how much money do they raise?

Grade F

2 a How many decimal places are there in each of these numbers?

  i 4.09   **ii** 32.609   **iii** 4.5

b Round these numbers to one decimal place.

  i 2.32   **ii** 6.08   **iii** 15.856

c Round these numbers to two decimal places.

  i 3.454   **ii** 16.089   **iii** 7.625

d Round these numbers to three decimal places.

  i 4.9743   **ii** 6.2159   **iii** 0.0076

Remember: You must revise all content from Grade G to the level that you are currently working at.

Top Tip!
Make sure the decimal points are lined up.

Adding and subtracting decimals

- When **adding** and **subtracting** decimals, you must use a **column** method.

- **Line up** the **decimal points** and use **zeros** to fill in any blanks.

- Do the addition or subtraction as normal, starting with the column on the right.

- The decimal point in the answer will be placed directly underneath the other decimal points.

Write 3.4 + 2.56 as:

$$\begin{array}{r} 3.40 \\ +\ 2.56 \\ \hline \end{array}$$

$$\begin{array}{r} 3.40 \\ +\ 2.56 \\ \hline 5.96 \end{array} \qquad \begin{array}{r} 6.\,{}^6\cancel{7}\,{}^1 0 \\ -\ 2.48 \\ \hline 4.22 \end{array}$$

F

Multiplying and dividing decimals by single-digit numbers

- Solve these problems in the same way as for normal short multiplication and division.

- As in addition and subtraction, the decimal point in the answer will be underneath the decimal point in the calculation.

Write 4.32 × 3 as:

$$\begin{array}{r} 4.32 \\ \times\ \ \ \ 3 \\ \hline 12.96 \end{array}$$

and 3.45 ÷ 3 as:

$$\begin{array}{r} 1.15 \\ 3\overline{)\ 3.4\,{}^15} \end{array}$$

F-E

Long multiplication with decimals

- Solve these problems in the same way as for **normal long multiplication**.

- As before, keep the decimal points **in line**.

3.26 × 24

$$\begin{array}{r} 3.26 \\ \times\ \ \ 24 \\ \hline 13\ \ 04 \\ 65\ \ 20 \\ \hline 78.24 \end{array}$$

E-D

Multiplying decimals

- Solve these problems in the same way as for **normal multiplications** but ignore the decimal points in the working.

- The number of decimal places in the answer will be the same as the total of **decimal places** in the numbers in the original calculation.

There are three decimal places in the numbers in the question so there will be three in the answer.

4.52 × 3.2

$$\begin{array}{r} 452 \\ \times\ \ \ 32 \\ \hline 904 \\ 1356\ 0 \\ \hline 1446\ 4 \end{array}$$

So **4.52** × **3.2** = 14.**464**.

E-D

Questions

Grade E

1 a Work out each of these.
- **i** 56.2 + 1.6
- **ii** 25.6 + 5.5
- **iii** 32.6 + 3.9
- **iv** 4.9 − 1.3
- **v** 8.43 − 2.6
- **vi** 28.6 − 14.9

b Work out each of these.
- **i** 3.5 × 4
- **ii** 4.7 × 3
- **iii** 3.6 × 3
- **iv** 14.45 ÷ 5
- **v** 15.06 ÷ 6
- **vi** 13.95 ÷ 3

c Work out each of these.
- **i** 4.61 × 23
- **ii** 4.52 × 41
- **iii** 1.72 × 34
- **iv** 3.3 × 0.4
- **v** 0.13 × 0.7
- **vi** 14.2 × 2.8

More fractions

Adding and subtracting fractions

- When adding and subtracting fractions you must use equivalent fractions to make the denominators of the fractions the same.

$$\frac{3}{8} + \frac{1}{5} = \frac{3 \times 5}{8 \times 5} + \frac{1 \times 8}{5 \times 8} = \frac{15 + 8}{40} = \frac{23}{40}$$

- When adding or subtracting mixed numbers, split the calculation into whole numbers and fractions.

$$4\frac{2}{5} - 1\frac{3}{4} = (4 - 1) + \frac{2}{5} - \frac{3}{4} = 3 + \frac{8}{20} - \frac{15}{20} = 3 + -\frac{7}{20} = 2\frac{13}{20}$$

Top Tip!

Notice that you may have to split one of the whole numbers, if the answer to the fractional part is negative.

Multiplying fractions

- To **multiply** two fractions, multiply the numerators and multiply the denominators.

$$\frac{3}{4} \times \frac{3}{7} = \frac{3 \times 3}{4 \times 7} = \frac{9}{28}$$

- When multiplying **mixed numbers**, change the mixed numbers into **top-heavy** fractions then multiply, as for ordinary fractions.

- **Cancel** any common factors in the top and bottom before multiplying.

- Convert the **final answer** back to a **mixed number** if necessary.

$$3\frac{1}{4} \times 1\frac{1}{5} = \frac{13}{4} \times \frac{6}{5} = \frac{13 \times 6^3}{4^2 \times 5} = \frac{39}{10} = 3\frac{9}{10}$$

Dividing fractions

- To **divide** by a fraction, turn it upside down and multiply by it.

$$\frac{3}{8} \div \frac{7}{9} = \frac{3}{8} \times \frac{9}{7} = \frac{27}{56}$$

- When dividing mixed numbers, change them into top-heavy fractions and then divide, as for ordinary fractions.

- When the second fraction has been turned upside down, cancel before multiplying.

$$1\frac{1}{4} \div 1\frac{7}{8} = \frac{5}{4} \div \frac{15}{8} = \frac{5^1}{4_1} \times \frac{8^2}{15_3} = \frac{2}{3}$$

Questions

Grade C

1 a Work out each of these.

i $\frac{1}{4} + \frac{3}{7}$

ii $\frac{5}{6} + \frac{4}{9}$

iii $3\frac{2}{3} + 2\frac{2}{5}$

b Work out each of these.

i $\frac{3}{5} - \frac{1}{6}$

ii $\frac{8}{9} - \frac{2}{3}$

iii $2\frac{1}{4} - 1\frac{2}{3}$

c Work out each of these.

i $\frac{3}{4} \times \frac{2}{9}$

ii $\frac{5}{8} \times \frac{4}{7}$

iii $1\frac{2}{5} \times 2\frac{3}{4}$

d Work out each of these.

i $\frac{3}{5} \div \frac{6}{7}$

ii $\frac{5}{6} \div \frac{10}{21}$

iii $3\frac{3}{5} \div 2\frac{1}{4}$

More number

Multiplying and dividing with negative numbers

- The rules for multiplying and dividing with negative numbers are:
  - when the signs of the numbers are the **same**, the answer is **positive**
  - when the signs of the numbers are **different**, the answer is **negative**.

$+3 \times -4 = -12$, $+12 \div +3 = +4$, $-15 \div -5 = +3$, $-4 \times +6 = -24$

Top Tip!

You do not have to write a plus sign in front of a positive number but you must put a minus sign in front of a negative number.

— E

Rounding to one significant figure

- **Significant figures** are the digits of a number, from the first to the last non-zero digit.

  3400 has two significant figures (2 sf), 0.06 has one significant figure (1 sf) and 67.45 has four significant figures (4 sf).

- To round a number to one significant figure (1 sf), just round the first two non-zero digits, then replace the rest of the digits in the number with zeros.

  432 is 400 to 1 sf, 0.087 is 0.9 to 1 sf and 35.9 is 40 to 1 sf.

Top Tip!

In GCSE exams you only have to round to one significant figure.

— D-C

Approximation of calculations

- To **approximate** the answer to a calculation, round the numbers in the calculation to one significant figure, then work out the approximate answer.

  38.2×9.6 can be rounded to 40×10 which is 400, so $38.2 \times 9.6 \approx 400$

  $48.3 \div 19.7$ rounds to $50 \div 20 = 2.5$, so $48.3 \div 19.7 \approx 2.5$

- The sign $\approx$ means 'approximately equal to'.

— D-C

Questions

Grade E

1 a Work out each of these.

   i $+3 \times -5$ **ii** -4×-6 **iii** $-7 \times +5$

b Work out each of these.

   i $+24 \div -6$ **ii** $-18 \div +9$ **iii** $-12 \div -4$

Grade D

2 a How many significant figures do these numbers have?

   i 6.8   **ii** 0.964   **iii** 120.8

b Round each of these numbers to one significant figure.

   i 3.8   **ii** 0.752   **iii** 58.7

Grade D

3 Find approximate answers to these calculations.

a 68.3×12.2

b $203.7 \div 38.1$

c $\dfrac{78.3 + 19.6}{21.8 - 9.8}$

d $\dfrac{42.1 \times 78.6}{4.7 \times 19.3}$

Ratio

Ratio

- A **ratio** is a way of **comparing** the sizes of two or more quantities.

- A **colon** (:) is used to show ratios.   3:4 and 6:20 are ratios.

- Quantities to be compared must be in the **same units** as ratio itself has no units.

- Ratios that have a common factor can be **cancelled** to give a ratio in its **simplest form**.

- Ratios can be expressed as **fractions**, where the **denominator** is the **sum** of the two parts of the ratio.

If a garden has lawn and flower beds in the ratio 3:4 then $\frac{3}{7}$ of the garden is lawn and $\frac{4}{7}$ is flower beds.

Top Tip!

The method for cancelling a ratio to its simplest form is the same as for cancelling a fraction. Look for the highest common factor.

Dividing amounts in ratios

- A quantity can be divided into **portions** that are in a certain **given ratio**.
- The process has three steps:
  - **add** the separate parts of the ratio
  - **divide** this number into the original quantity
  - **multiply** this answer by the original parts of the ratio.

To share £40 in the ratio 2:3: Add 2 + 3 = 5. Divide 40 ÷ 5 = 8. Multiply each part by 8.

2 × 8 = 16, 3 × 8 = 24. So £40 divided in the ratio 2:3 gives shares of £16 and £24.

Top Tip!

Always check that the two parts into which you have divided a quantity add up to the original amount.

Calculating with ratios

- When one part of a ratio is known, it is possible to calculate other values.
- The process has two steps:
  - use the given information to find a **unit value**
  - use the unit value to find the required information.

When the cost of a meal was shared between two families in the ratio 3:5 the smaller share was £22.50. How much did the meal cost altogether?

$\frac{3}{8}$ of the cost was £22.50, so $\frac{1}{8}$ was £7.50. The total cost was 8 × 7.5 = £60.

Questions

(Grade D)

1 a Express each of these as a ratio in its simplest form.

  i 12:36   **ii** 25:30   **iii** 18:30

  b Express each of these as a ratio in its simplest form (remember to express both parts in common units).

  i 50p to £3   **ii** 2 hours to 15 minutes
  iii 40 cm to 1 metre

(Grade C)

2 Divide the following amounts in the given ratios.

  a £500 in the ratio 1:4

  b 300 grams in the ratio 1:5
  c £400 in the ratio 3:5
  d 240 kg in the ratio 1:2

(Grade C)

3 a A catering box of crisps has two flavours, plain and beef, in the ratio 3:4. There are 42 packets of plain crisps. How many packets of beef crisps are there in the box?

  b The ratio of male teachers to female teachers in a school is 3:7. If there are 21 female teachers, how many teachers are there in total?

Speed and proportion

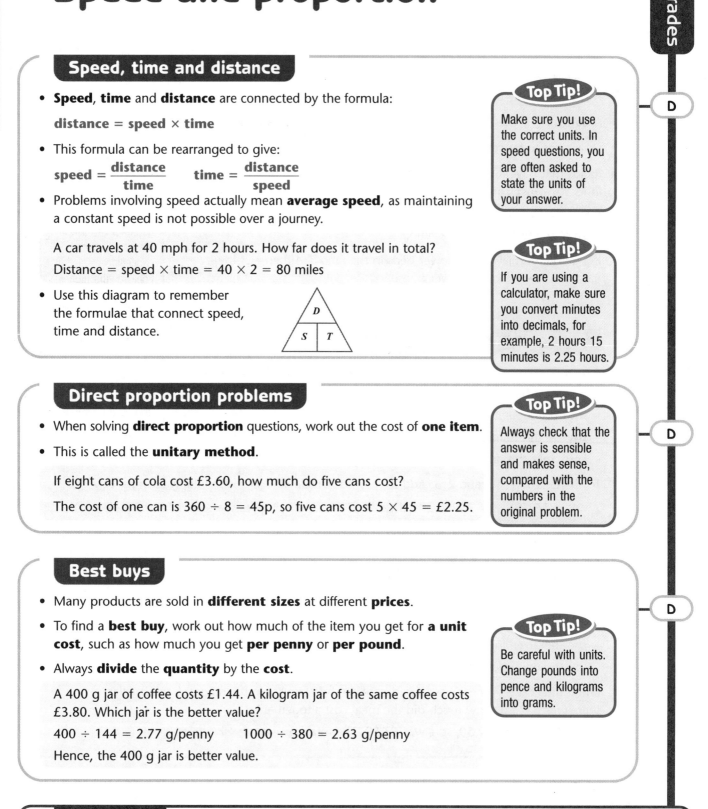

Speed, time and distance

- **Speed**, **time** and **distance** are connected by the formula:

  distance = speed × time

- This formula can be rearranged to give:

  $$speed = \frac{distance}{time} \qquad time = \frac{distance}{speed}$$

- Problems involving speed actually mean **average speed**, as maintaining a constant speed is not possible over a journey.

  A car travels at 40 mph for 2 hours. How far does it travel in total?

  Distance = speed × time = 40 × 2 = 80 miles

- Use this diagram to remember the formulae that connect speed, time and distance.

> **Top Tip!**
> Make sure you use the correct units. In speed questions, you are often asked to state the units of your answer.

> **Top Tip!**
> If you are using a calculator, make sure you convert minutes into decimals, for example, 2 hours 15 minutes is 2.25 hours.

D

Direct proportion problems

- When solving **direct proportion** questions, work out the cost of **one item**.

- This is called the **unitary method**.

  If eight cans of cola cost £3.60, how much do five cans cost?

  The cost of one can is 360 ÷ 8 = 45p, so five cans cost 5 × 45 = £2.25.

> **Top Tip!**
> Always check that the answer is sensible and makes sense, compared with the numbers in the original problem.

D

Best buys

- Many products are sold in **different sizes** at different **prices**.

- To find a **best buy**, work out how much of the item you get for **a unit cost**, such as how much you get **per penny** or **per pound**.

- Always **divide** the **quantity** by the **cost**.

  A 400 g jar of coffee costs £1.44. A kilogram jar of the same coffee costs £3.80. Which jar is the better value?

  400 ÷ 144 = 2.77 g/penny   1000 ÷ 380 = 2.63 g/penny

  Hence, the 400 g jar is better value.

> **Top Tip!**
> Be careful with units. Change pounds into pence and kilograms into grams.

D

Questions

Grade D

1 a A motorist travels a distance of 75 miles in 2 hours. What is the average speed?

b A cyclist travels for $3\frac{1}{2}$ hours at an average speed of 15 km per hour. How far has she travelled?

Grade C

2 a 40 bricks weigh 50 kg. How much will 25 bricks weigh?

b How many bricks are there on a pallet weighing 200 kg, if you ignore the weight of the pallet?

Grade C

3 a A large tube of toothpaste contains 250 g and costs £1.80. A travel-size tube contains 75 g and costs 52p. Which is the better value?

b Which is the better mark:

    62 out of 80
    95 out of 120?

Percentages

Equivalent percentages, fractions and decimals

- **Fractions**, **percentages** and **decimals** are all different ways of expressing parts of a whole.

- This table shows how to convert from one to another.

	Decimal	Fraction	Percentage
Percentage to:	Divide by 100 (or move the decimal point two places left). $60\% = 0.6$	Put it over 100 and cancel if possible. $55\% = \frac{55}{100} = \frac{11}{20}$	
Fraction to:	Divide the numerator by the denominator. $\frac{4}{5} = 4 \div 5 = 0.8$		Divide the numerator by the denominator and multiply by 100. $\frac{7}{8} = 7 \div 8 \times 100$ $= 87.5\%$
Decimal to:		Make the denominator 10 or 100 then cancel if possible. $0.68 = \frac{68}{100} = \frac{17}{25}$	Multiply by 100 (or move the decimal point two places right). $0.76 = 76\%$

Top Tip!

Most of the questions in GCSE use very basic values so it is worth learning some of them, such as $0.1 = \frac{1}{10} = 10\%$, $0.25 = \frac{1}{4} = 25\%$.

The percentage multiplier

- Using the **percentage multiplier** is the best way to solve percentage problems.

- The percentage multiplier is the **percentage** expressed as a **decimal**.

    72% gives a multiplier of 0.72, 20% is a multiplier of 0.20 or 0.2.

- The multiplier for a percentage **increase** or **decrease** is the percentage multiplier **added to 1** or **subtracted from 1**.

    An 8% increase is a multiplier of 1.08 (1 + 0.08), a 5% decrease is a multiplier of 0.95 (1 – 0.05).

Top Tip!

Learn how to use multipliers as they make percentage calculations easier and more accurate.

Questions

Grade D

1 a Write the following percentages as decimals.
   i 30%   **ii** 88%

b Write the following percentages as fractions.
   i 90%   **ii** 32%

c Write the following decimals as percentages.
   i 0.85   **ii** 0.15

d Write the following decimals as fractions.
   i 0.8   **ii** 0.08

e Write the following fractions as decimals.
   i $\frac{5}{8}$ **ii** $\frac{7}{20}$

f Write the following fractions as percentages.
   i $\frac{9}{25}$ **ii** $\frac{1}{20}$

Grade D

2 a Write down the percentage multiplier for each of these.
   i 80%   **ii** 7%   **iii** 22%

b Write down the multiplier for each percentage increase.
   i 5%   **ii** 12%   **iii** 3.2%

c Write down the multiplier for each percentage decrease.
   i 8%   **ii** 15%   **iii** 4%

Calculating a percentage of a quantity

- To calculate a percentage of a quantity simply **multiply** the quantity by the **percentage multiplier**.

  What is 8% of 56 kg?

  Work it out as $0.08 \times 56 = 4.48$ kg.

- On the non-calculator paper, percentages will be **based on 10%**.

- To work out 10%, just **divide the quantity by 10**.

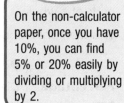
Top Tip!

On the non-calculator paper, once you have 10%, you can find 5% or 20% easily by dividing or multiplying by 2.

E

Percentage increase or decrease

- To calculate the new value after a quantity is **increased** or **decreased** by a percentage, simply **multiply** the original quantity by the **percentage multiplier** for the increase or decrease.

  What is the new cost after a price of £56 is decreased by 15%?

  Work it out as $0.85 \times 56 = £47.60$.

  Jamil gets a wage increase of 4%. His previous wage was £180 per week. How much does he get now?

  Work it out as $180 \times 1.04 = £187.20$.

Top Tip!

If the calculator shows a number such as 47.6 as the answer to a money problem, always put the extra zero into the answer, so you write this down as £47.60.

D

One quantity as a percentage of another

- To calculate one **quantity** as **a percentage** of **another** just divide the first quantity by the second. This will give a decimal, which can be converted to a percentage.

  A plant grows from 30 cm to 39 cm in a week. What is the percentage growth?

  The increase is 9 cm, $9 \div 30 = 0.3$ and this is 30%.

Top Tip!

Always divide by the original quantity, otherwise you will get no marks.

C

Questions

Grade E

1 a Work these out.

   i 15% of £70   **ii** 32% of 60 kg

b A jacket is priced at £60. Its price is reduced by 12% in a sale.

   i What is 12% of £60?

   ii What is the sale price of the jacket?

Grade D

2 a Increase £150 by 12%.

b Decrease 72 kg by 8%.

Grade C

3 a The average attendance at Barnsley football club in 2005 was 14 800.

In 2006 it was 15 540.

   i By how much had the average attendance gone up?

   ii What is the percentage increase in attendance?

b After her diet, Carol's weight had gone from 80 kg to 64 kg.

What is the percentage decrease in her weight?

Number checklist

I can...

☐ recall the times tables up to 10×10

☐ use BODMAS to do calculations in the correct order

☐ identify the place value of digits in whole numbers

☐ round numbers to the nearest 10 or 100

☐ add and subtract numbers with up to four digits without a calculator

☐ multiply numbers by a single-digit number

☐ state what fraction of a shape is shaded

☐ shade in a fraction of a shape

☐ add and subtract simple fractions with the same denominator

☐ recognise equivalent fractions

☐ cancel a fraction

☐ change top-heavy fractions into mixed numbers and vice versa

☐ find a fraction of an integer

☐ recognise the multiples of the first 10 whole numbers

☐ recognise square numbers up to 100

☐ find equivalent fractions, decimals and percentages

☐ recognise that a number on the left on a number line is smaller than a number on the right

☐ read negative numbers on scales such as thermometers

You are working at (Grade G) level.

☐ divide numbers by a single-digit number

☐ put fractions in order of size

☐ add fractions with different denominators

☐ solve fraction problems expressed in words

☐ compare fractions of quantities

☐ find factors of numbers less than 100

☐ add and subtract with negative numbers

☐ write down the squares of numbers up to 15×15

☐ write down the cubes of 1, 2, 3, 4, 5 and 10

☐ use a calculator to find square roots

☐ do long multiplication

☐ do long division

☐ solve real-life problems involving multiplication and division

☐ round decimal numbers to one, two or three decimal places

☐ find percentages of a quantity

☐ change mixed numbers into top-heavy fractions

You are working at (Grade F) level.

- [ ] multiply fractions
- [ ] add and subtract mixed numbers
- [ ] calculate powers of numbers
- [ ] recognise prime numbers under 100
- [ ] use the four rules with decimals
- [ ] change decimals to fractions
- [ ] change fractions to decimals
- [ ] simplify a ratio
- [ ] find a percentage of any quantity

You are working at (Grade E) level.

- [ ] work out one quantity as a fraction of another
- [ ] solve problems using negative numbers
- [ ] multiply and divide by powers of 10
- [ ] multiply together numbers that are multiples of powers of 10
- [ ] round numbers to one significant figure
- [ ] estimate the answer to a calculation
- [ ] order lists of numbers containing decimals, fractions and percentages
- [ ] multiply and divide fractions
- [ ] calculate with speed, distance and time
- [ ] compare prices to find 'best buys'
- [ ] find the new value after a percentage increase or decrease
- [ ] find one quantity as a percentage of another
- [ ] solve problems involving simple negative numbers
- [ ] multiply and divide fractions

You are working at (Grade D) level.

- [ ] work out a reciprocal
- [ ] recognise and work out terminating and recurring decimals
- [ ] write a number as a product of prime factors
- [ ] use the index laws to simplify calculations and expressions
- [ ] multiply and divide with negative numbers
- [ ] multiply and divide with mixed numbers
- [ ] find a percentage increase
- [ ] work out the LCM and HCF of two numbers
- [ ] solve problems using ratio in appropriate situations

You are working at (Grade C) level.

Perimeter and area

Perimeter

- The **perimeter** of a shape is the **sum** of the lengths of all its **sides**.

The perimeter of this rectangle is 22 cm.

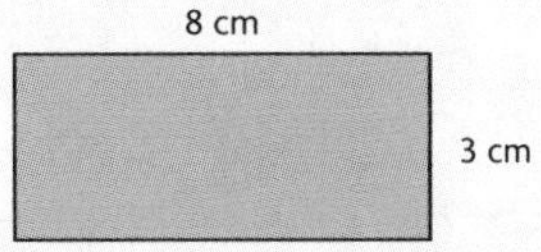

8 cm

3 cm

- A **compound shape** is a 2-D shape that is made up of other simple shapes, such as rectangles and triangles.

These are compound shapes.

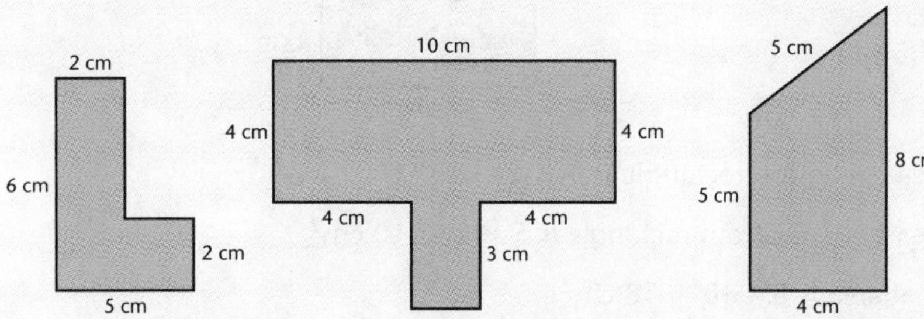

2 cm
6 cm
2 cm
5 cm

10 cm
4 cm
4 cm
4 cm
4 cm
3 cm

5 cm
8 cm
5 cm
4 cm

Top Tip!

When working with compound shapes, draw lines to split them into simple shapes such as rectangles and triangles, then write down the dimensions of the new shapes.

Area of irregular shapes (counting squares)

- The basic **units** of area are square centimetres (**cm²**), square metres (**m²**) and square kilometres (**km²**).

- To work out the area of an irregular shape, trace it onto a sheet of **centimetre squares**, say, and then **count** the squares.

These shapes are drawn on a centimetre grid.

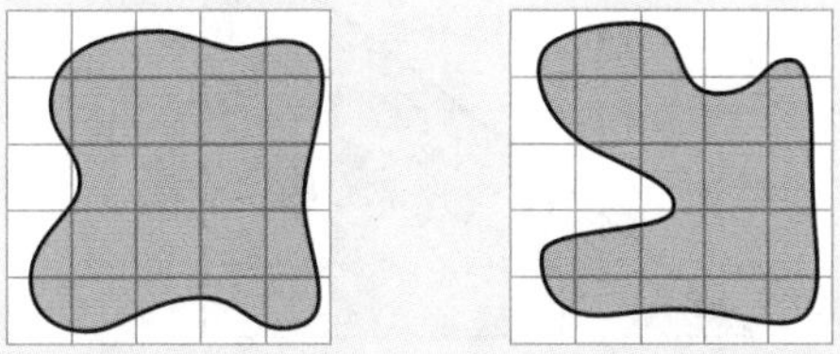

Top Tip!

In exams, the grid is always drawn for any irregular shapes so all you have to do is count whole and part squares.

Area of a rectangle

- The area of a rectangle can be calculated by the formula:

area = length × width $A = lw$

The area of the rectangle above is:

$A = 8 \times 3 = 24$ cm²

Top Tip!

Take care with units. In area questions you are often asked to 'State the units of your answer.'

Questions

Grade G

1 Find the perimeter of each of the compound shapes shown above.

Grade G

2 By counting squares, find the area of the two irregular shapes above.

Grade F

3 **a** Find the area of a rectangle with sides of 6 m and 2.5 m.

 b Find the area of a rectangle with sides of 3.2 cm and 5 cm.

Remember: You must revise all content from Grade G to the level that you are currently working at.

Area of a compound shape

D

- To find the area of a compound shape, **divide** the shape into regular shapes, such as rectangles and triangles, and calculate the areas of the separate parts.

> **Top Tip!**
> The sides of the rectangles into which you split the compound shape will probably need to be calculated by subtraction.

Find the area of this shape.

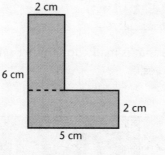

The area of the top rectangle is $4 \times 2 = 8$ cm^2

The area of the bottom rectangle is $5 \times 2 = 10$ cm^2

The total area is $8 + 10 = 18$ cm^2

Area of a triangle

E-D

- To work out the area of a triangle use the formula:

area = $\frac{1}{2}$ × base × height

$A = \frac{1}{2} bh$

Find the area of this triangle.

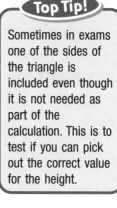

> **Top Tip!**
> Sometimes in exams one of the sides of the triangle is included even though it is not needed as part of the calculation. This is to test if you can pick out the correct value for the height.

$A = \frac{1}{2} \times 8 \times 6 = 4 \times 6 = 24$ cm^2

- The height of a triangle is the **perpendicular distance** from the base to the top point or **vertex**.

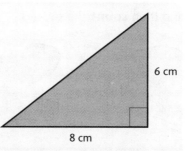

Questions

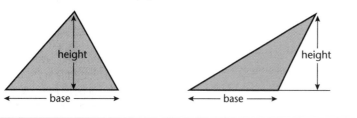

Grade D

1 Find the area of a triangle with:
  a base 7 cm, height 10 cm
  b base 5 m, height 5 m.

Grade D

2 Find the area of each of these two compound shapes.
  a
  b

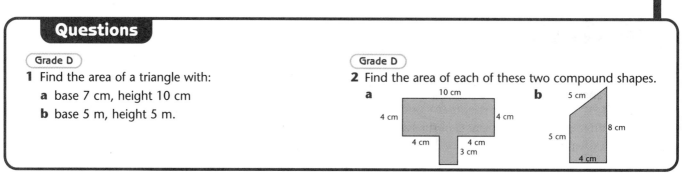

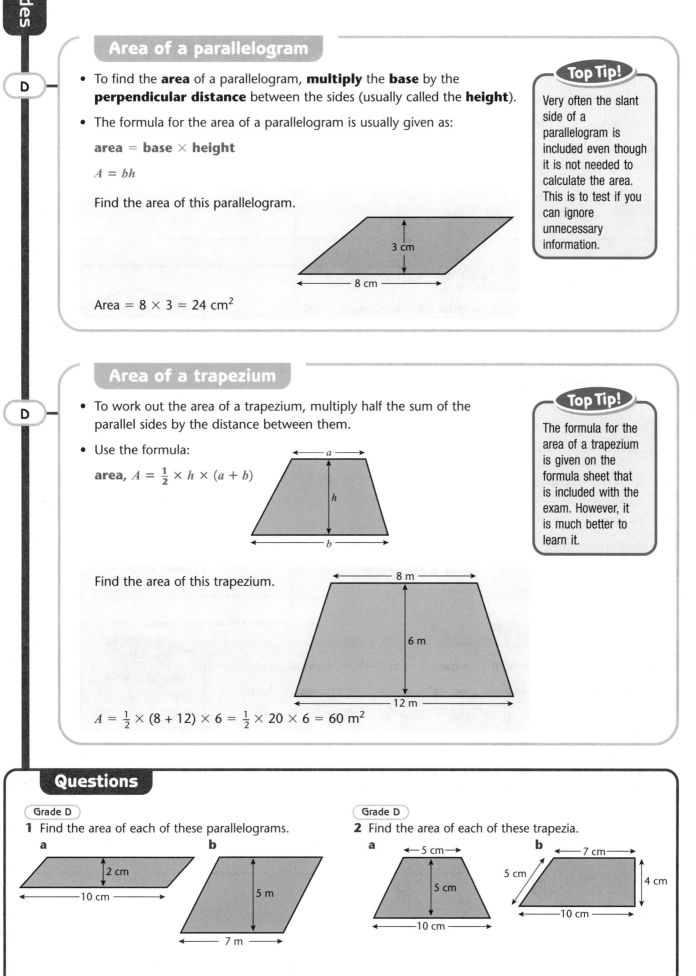

D

Area of a parallelogram

- To find the **area** of a parallelogram, **multiply** the **base** by the **perpendicular distance** between the sides (usually called the **height**).

- The formula for the area of a parallelogram is usually given as:

area = **base** × **height**

$A = bh$

Find the area of this parallelogram.

3 cm

8 cm

Area = 8 × 3 = 24 cm²

Top Tip!

Very often the slant side of a parallelogram is included even though it is not needed to calculate the area. This is to test if you can ignore unnecessary information.

D

Area of a trapezium

- To work out the area of a trapezium, multiply half the sum of the parallel sides by the distance between them.

- Use the formula:

area, $A = \frac{1}{2} \times h \times (a + b)$

a

h

b

Top Tip!

The formula for the area of a trapezium is given on the formula sheet that is included with the exam. However, it is much better to learn it.

Find the area of this trapezium.

8 m

6 m

12 m

$A = \frac{1}{2} \times (8 + 12) \times 6 = \frac{1}{2} \times 20 \times 6 = 60$ m²

Questions

Grade D

1 Find the area of each of these parallelograms.

a

2 cm

10 cm

b

5 m

7 m

Grade D

2 Find the area of each of these trapezia.

a

5 cm

5 cm

10 cm

b

7 cm

5 cm

4 cm

10 cm

Dimensional analysis

Dimensional analysis

- The **dimensions** of a shape are the **lengths** of the sides and the sizes of the **angles**.

- By looking at a **formula** it is possible to tell if it is a length, area or volume.

- You already know many formulae for lengths, areas and volumes:
  - the perimeter of a rectangle $P = 2l + 2w$
  - the area of a square $A = x^2$
  - the volume of a cuboid $V = lwh$

C

Dimensions of length

- All formulae that represent **lengths** are **one-dimensional**.

- All formulae that represent **lengths** have **single letters** in them.

- Length formulae include:
  - the circumference of a circle $C = 2\pi r$
  - the perimeter of a square $P = 4x$
  - the perimeter of a triangle $a + b + c$

> **Top Tip!**
> π is a number, not a letter, so look for letters on their own in length formula.

C

Dimensions of area

- All formulae that represent **areas** are **two-dimensional**.

- All formulae that represent **areas** have letters in '**pairs**' in them.

- Area formulae include: $A = \pi r^2$, $A = bh$, $ac + b^2$

> **Top Tip!**
> A term such as r^2 means $r \times r$ so it is a 'pair' of letters.

C

Dimensions of volume

- All formulae that represent **volumes** are **three-dimensional**.

- All formulae that represent **volumes** have letters in '**threes**' in them.

- Volume formulae include: $V = x^3$, $V = lbh$, $abc + ab^2$

So x, y and z represent lengths.

> **Top Tip!**
> A term such as r^3 means $r \times r \times r$ so it is 'three' letters together.

Here are three expressions. One of them is a length, one of them is an area and one of them is a volume. Which is which?

$xy + xz + yz$ xyz $x + y + z$

In order, the expressions are: area, volume and length.

C

Questions

Grade C

1 Three of the following formulae are lengths, three of them are areas and three of them are volumes. Identify which is which.

   a πr **b** $x^2 y$ **c** pq **d** pqr **e** $a + b$ **f** $\pi r^2 h$ **g** $\frac{1}{2}bh$ **h** $2p + 2q$ **i** $p^2 + q^2$

Symmetry

G-F

Lines of symmetry

- Many two-dimensional shapes have one or more **lines of symmetry**.

- A **line of symmetry** is a line that can be drawn through a shape so that what can be seen on one side of the line is the **mirror image** of what is on the other side.

- Lines of symmetry are also called **mirror lines**.

These shapes all have lines of symmetry.

Top Tip!

The easiest way to check a shape for a line of symmetry is to trace it and fold the tracing paper along the mirror line. The object and its image will be on top of each other.

F

Rotational symmetry

- A two-dimensional shape has **rotational symmetry** if it can be **rotated** about a **point** to look exactly the same in a new position.

- The **order** of rotational symmetry is the **number of different positions** in which the shape looks the **same** when it is rotated.

These shapes all have rotational symmetry.

Top Tip!

The easiest way to check for rotational symmetry is to trace the shape and, as you rotate the tracing paper a full turn about the centre, count how many times it looks the same.

D

Planes of symmetry

- Many two-dimensional shapes have one or more **planes of symmetry**.

- A **plane of symmetry** is a flat surface that can be cut through a shape so that what is on one side of the plane is the **mirror image** of what is on the other side.

These solids all have planes of symmetry.

Top Tip!

Imagine the solids are made of modelling clay and you are cutting them with a knife.

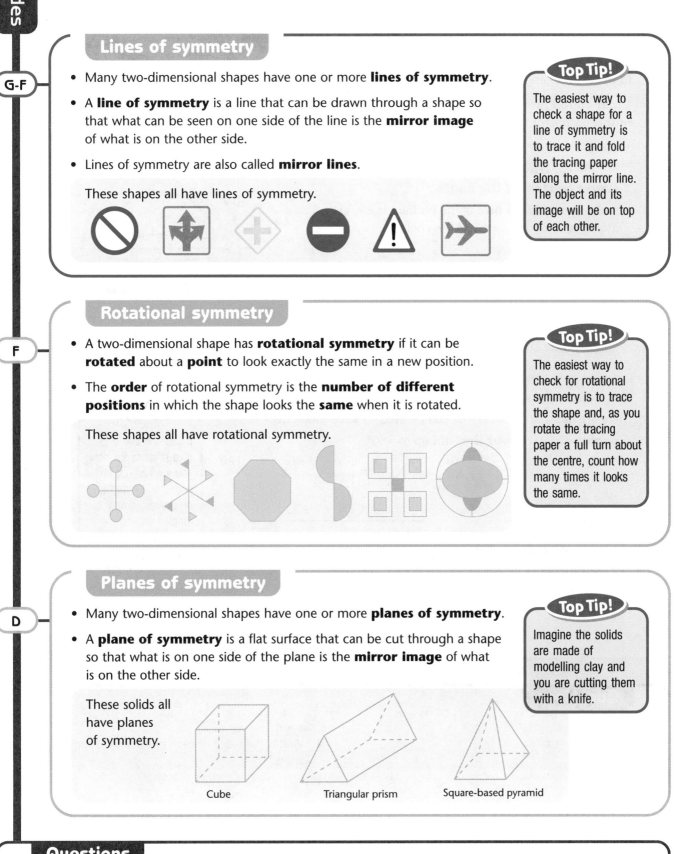

Cube   Triangular prism   Square-based pyramid

Questions

Grade D

1 Use tracing paper or otherwise to find out how many lines of symmetry the plane shapes, above (top), have.

Grade D

2 Use tracing paper or otherwise to find out the order of rotational symmetry the plane shapes, above (middle), have.

Grade D

3 How many planes of symmetry do the solids shown above have?

Angles

Measuring and drawing angles

- To measure an angle, use a **protractor**.

- A **half round** protractor measures up to 180° and a **full round** protractor measures up to 360°.

- It is important to remember two rules when measuring angles:

  - put the **centre of the protractor** exactly over the **centre (vertex) of the angle**

  - make sure the **zero line** of the protractor is exactly **along one of the arms** of the angle.

> **Top Tip!**
> Full round protractors are best, especially for working with bearings, which come later.

To measure reflex angles, such as GHI shown below, it is easier to use a circular protractor.

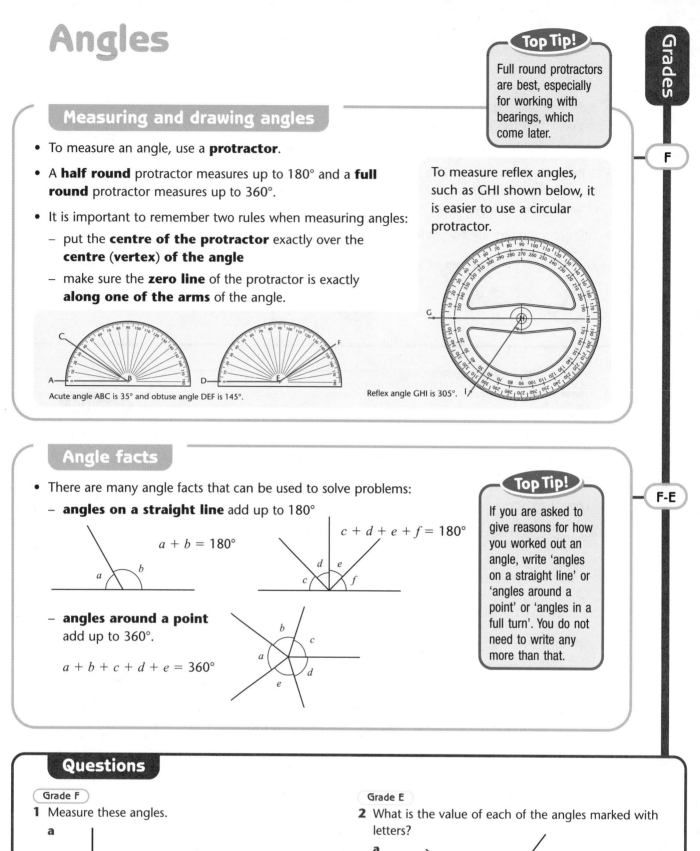

Acute angle ABC is 35° and obtuse angle DEF is 145°.

Reflex angle GHI is 305°.

Angle facts

- There are many angle facts that can be used to solve problems:

  - **angles on a straight line** add up to 180°

$$a + b = 180°$$

$$c + d + e + f = 180°$$

  - **angles around a point** add up to 360°.

$$a + b + c + d + e = 360°$$

> **Top Tip!**
> If you are asked to give reasons for how you worked out an angle, write 'angles on a straight line' or 'angles around a point' or 'angles in a full turn'. You do not need to write any more than that.

Questions

Grade F

1 Measure these angles.

a

b

Grade E

2 What is the value of each of the angles marked with letters?

a

a 52°

b

81°

b

73°

Angles in a triangle

E–D

- The **angles in a triangle** add up to **180°**.
- There are four types of triangle.

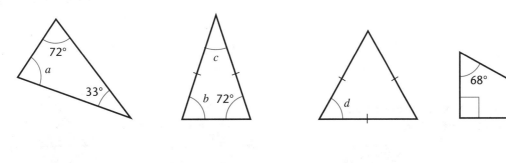

A **scalene** triangle doesn't have any two angles the same or any two sides the same.

An **isosceles** triangle has two angles the same and two sides the same.

An **equilateral triangle** has all three angles the same (60°) and all three sides the same length.

A **right-angled** triangle has one right angle (90°).

Angles in a quadrilateral

E–D

- The angles in a **quadrilateral** add up to **360°**.
- Any quadrilateral can be formed from **two triangles**.

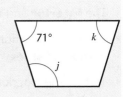

Some quadrilaterals are shown below.

Square

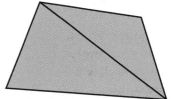

Rectangle

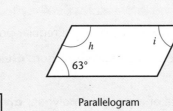

Parallelogram

Isosceles trapezium

Questions

Grade E

1 Find the values of the angles marked with letters in the triangles, above.

Grade D

2 Use the diagram, above, of the quadrilateral with one diagonal to explain why the angles in a quadrilateral add up to 360°.

Grade D

3 Find the values of the angles marked with letters in the quadrilaterals, above.

Polygons

Polygons

- The angles in a **triangle** add up to **180°**.
- The angles in a **quadrilateral** add up to **360°**.
- The angles in a **pentagon** add up to **540°**.
- The angles in a **hexagon** add up to **720°**.
- The angle sum increases by 180° for each side added because each time an extra triangle is added.
- In **any polygon** there are **two fewer triangles** than the **number of sides**.

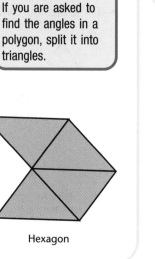

Top Tip!

If you are asked to find the angles in a polygon, split it into triangles.

— E-D

Triangle   Quadrilateral   Pentagon   Hexagon

Regular polygons

- A **regular polygon** is a polygon with all its sides the same length.
- In any regular polygon, all the **interior angles**, i, are equal, and all of the **exterior angles**, e, are equal.

Here are three regular polygons.

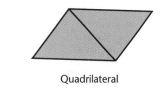

C

Interior and exterior angles in a regular polygon

- The **interior** and **exterior** angles of some regular polygons are shown below.
- To find the interior angle of any regular polygon, **divide the angle sum** by the number of **sides**.
- To find the exterior angle of any regular polygon, **divide 360°** by the number of **sides**.

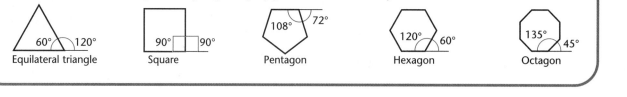

Equilateral triangle   Square   Pentagon   Hexagon   Octagon

C

Questions

Grade C

1 What is the interior angle of a regular: **a** octagon   **b** nonagon?

Grade C

2 What is the exterior angle of a regular: **a** octagon   **b** nonagon?

Grade C

3 What is the connection between the interior and exterior angles of any regular polygon?

Parallel lines and angles

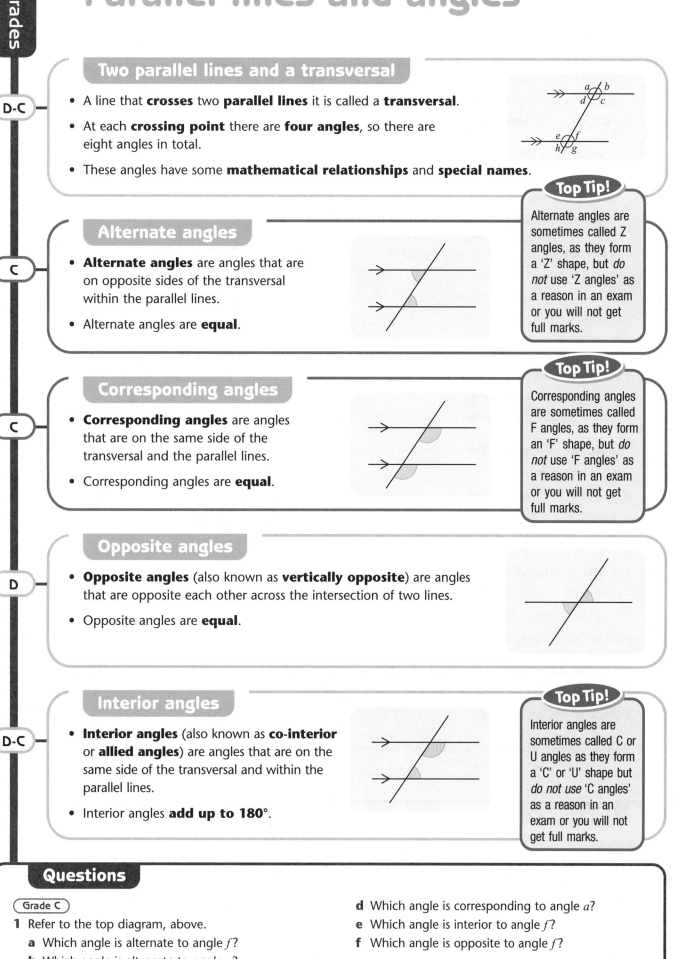

D-C

Two parallel lines and a transversal

- A line that **crosses** two **parallel lines** it is called a **transversal**.

- At each **crossing point** there are **four angles**, so there are eight angles in total.

- These angles have some **mathematical relationships** and **special names**.

C

Alternate angles

- **Alternate angles** are angles that are on opposite sides of the transversal within the parallel lines.

- Alternate angles are **equal**.

> **Top Tip!**
>
> Alternate angles are sometimes called Z angles, as they form a 'Z' shape, but *do not* use 'Z angles' as a reason in an exam or you will not get full marks.

C

Corresponding angles

- **Corresponding angles** are angles that are on the same side of the transversal and the parallel lines.

- Corresponding angles are **equal**.

> **Top Tip!**
>
> Corresponding angles are sometimes called F angles, as they form an 'F' shape, but *do not* use 'F angles' as a reason in an exam or you will not get full marks.

D

Opposite angles

- **Opposite angles** (also known as **vertically opposite**) are angles that are opposite each other across the intersection of two lines.

- Opposite angles are **equal**.

D-C

Interior angles

- **Interior angles** (also known as **co-interior** or **allied angles**) are angles that are on the same side of the transversal and within the parallel lines.

- Interior angles **add up to 180°**.

> **Top Tip!**
>
> Interior angles are sometimes called C or U angles as they form a 'C' or 'U' shape but *do not use* 'C angles' as a reason in an exam or you will not get full marks.

Questions

(Grade C)

1 Refer to the top diagram, above.
  a Which angle is alternate to angle f?
  b Which angle is alternate to angle c?
  c Which angle is corresponding to angle f?
  d Which angle is corresponding to angle a?
  e Which angle is interior to angle f?
  f Which angle is opposite to angle f?

Quadrilaterals

Special quadrilaterals

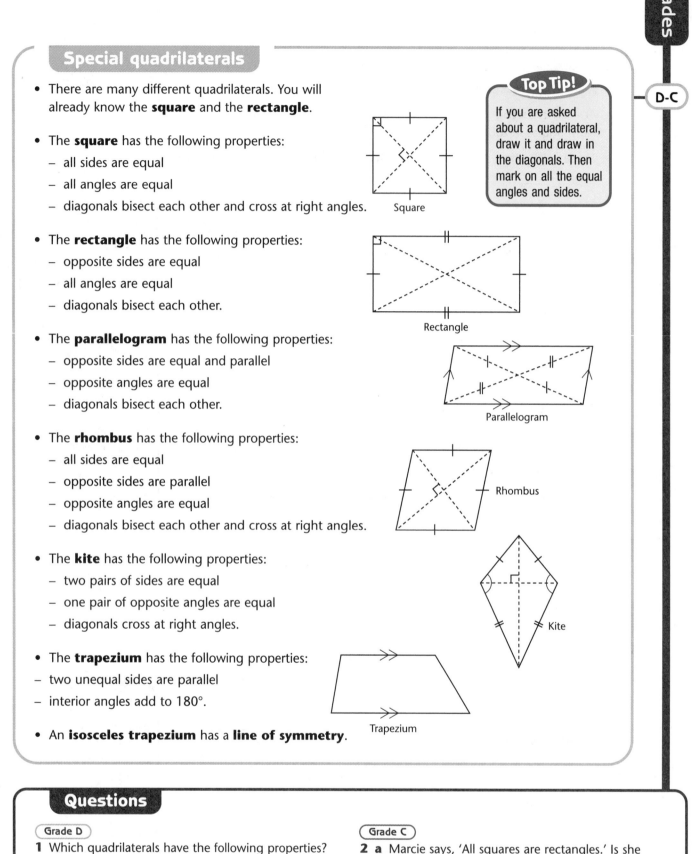

- There are many different quadrilaterals. You will already know the **square** and the **rectangle**.

Top Tip!

If you are asked about a quadrilateral, draw it and draw in the diagonals. Then mark on all the equal angles and sides.

- The **square** has the following properties:
  - all sides are equal
  - all angles are equal
  - diagonals bisect each other and cross at right angles.

Square

- The **rectangle** has the following properties:
  - opposite sides are equal
  - all angles are equal
  - diagonals bisect each other.

Rectangle

- The **parallelogram** has the following properties:
  - opposite sides are equal and parallel
  - opposite angles are equal
  - diagonals bisect each other.

Parallelogram

- The **rhombus** has the following properties:
  - all sides are equal
  - opposite sides are parallel
  - opposite angles are equal
  - diagonals bisect each other and cross at right angles.

Rhombus

- The **kite** has the following properties:
  - two pairs of sides are equal
  - one pair of opposite angles are equal
  - diagonals cross at right angles.

Kite

- The **trapezium** has the following properties:
- two unequal sides are parallel
- interior angles add to 180°.

Trapezium

- An **isosceles trapezium** has a **line of symmetry**.

Questions

Grade D

1 Which quadrilaterals have the following properties?
  a Diagonals cross at right angles.
  b Diagonals bisect each other.
  c All sides are equal.
  d Both pairs of opposite sides are equal.
  e Both pairs of opposite angles are equal.
  f One pair of sides are parallel.
  g One pair of angles are equal.

Grade C

2 a Marcie says, 'All squares are rectangles.' Is she correct?
  b Milly says, 'All rhombuses are parallelograms.' Is she correct?
  c Molly says, 'All kites are rhombuses.' Is she correct?

Bearings

D

Bearings

- **Bearings** are used to describe the position and direction of one object in relation to another object.

- Bearings are used to describe positions of aircraft or ships, for example. Hikers also use bearings to make sure they do not get lost in bad weather.

- The **bearing** of an object, from where you are standing, is the angle through which you turn towards the object, in a clockwise direction, as you turn from facing north.

- Bearings are also known as **three-figure bearings** as it is normal to give a bearing of 60° as 060°.

- The main points of the **compass rose** have bearings of 000°, 045°, 090°, etc.

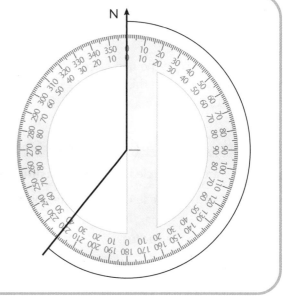

D

Measuring a bearing

- Draw a north line.
  - Place the centre of the protractor exactly on the point from which you are measuring.
  - Line up the 0° line with north.
  - Read the bearing, using the clockwise scale.

Questions

1 What bearing is being measured by the protractor, above?

2 a What is the **opposite** bearing to east?

   b What is the **opposite** bearing to north-east?

3 A plane is flying on a bearing of 315°. In what compass direction is it flying?

4 On a flat desert surface a man walks 1 kilometre north, then 1 kilometre west.
He then heads directly back to where he started.
On what bearing will he be walking?

Circles

Circles

- You need to know the terms, labelled on the diagram, that relate to **circles**.

- You also need to know about π (pronounced '**pi**'), a special number used in work with circles.

- The value of π is a **decimal** that goes on **for ever** but it is taken as 3.142 to three decimal places.

- All calculators should have a π **button** that will give the value as 3.141 592 654.

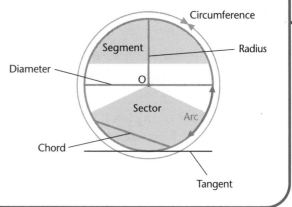

Circumference of a circle

- The **circumference** of a circle is the distance around the circle (the **perimeter**).

- The circumference of a circle is given by the formula:

  $C = \pi d$ or $C = 2\pi r$

  where d is the **diameter** and r is the **radius**.

- The formula you use depends on whether you are given the radius or the diameter.

  What is the circumference of a circle with a radius of 20 cm?

  $C = 2 \times \pi \times 20 = 40\pi = 125.7$ cm

- Sometimes on the non-calculator paper you are asked to give an answer in terms of π. In this case leave your answer as, for example, 40π.

> **Top Tip!**
>
> You can learn just one formula because you can always find the radius from the diameter, or vice versa, as: $d = 2r$

> **Top Tip!**
>
> Unless you are asked to give your answer in terms of π, use a calculator to work out circumferences and areas of circles and give answers correct to at least one decimal place.

Area of a circle

- The area of a circle is the **space inside** the circle.

- The area of a circle is given by the formula:

  $A = \pi r^2$

  where r is the **radius**.

- You must use the radius when calculating the area.

Questions

Grade D

1 a Calculate the circumference of a circle with radius 7 cm. Give your answer to one decimal place.

  b Calculate the circumference of a circle with diameter 12 cm. Give your answer to one decimal place.

  c Calculate the circumference of a circle with radius 4 cm. Leave your answer in terms of π.

Grade D

2 a Calculate the area of the circle, above, with diameter 8 cm.
Give your answer to one decimal place.

  b Calculate the area of a circle with radius 15 cm. Give your answer to one decimal place.

  c Calculate the area of a circle with radius 3 cm. Leave your answer in terms of π.

Scales

Scales

- You will come across **scales** in many places in everyday life, for example, on thermometers, car speedometers and on kitchen scales.

- When you read a scale, make sure you know what each **division** on the scale represents.

- When you read a scale, make sure that you read it in the **right direction**.

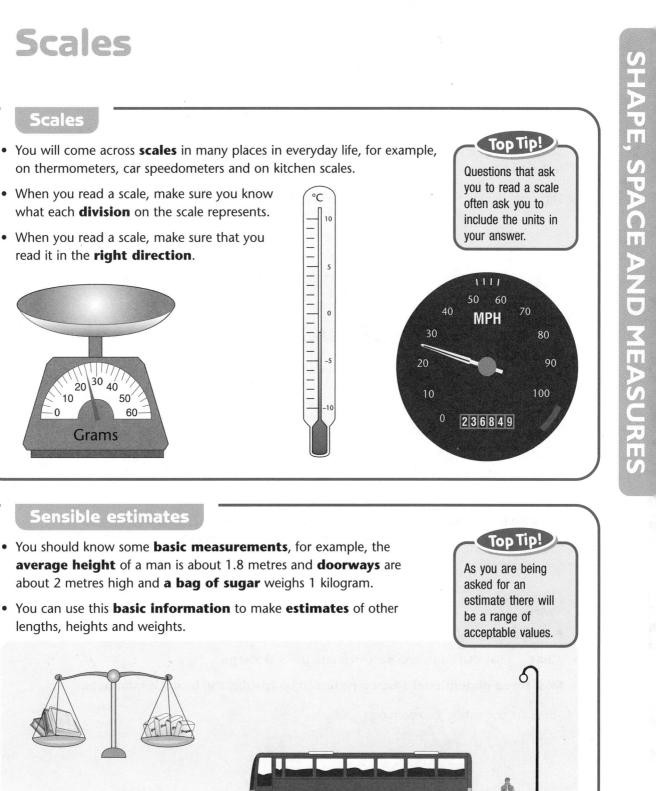

Top Tip!

Questions that ask you to read a scale often ask you to include the units in your answer.

Sensible estimates

- You should know some **basic measurements**, for example, the **average height** of a man is about 1.8 metres and **doorways** are about 2 metres high and **a bag of sugar** weighs 1 kilogram.

- You can use this **basic information** to make **estimates** of other lengths, heights and weights.

Top Tip!

As you are being asked for an estimate there will be a range of acceptable values.

Questions

(Grade G)

1 Include the units of your answer.

  a What temperature is being shown on the thermometer, above?

  b What speed is being shown on the car speedometer, above?

  c What weight is being shown on the kitchen scales, above?

(Grade G)

2 a From the picture, above, estimate:

    i the height of the lamppost

    ii the length of the bus.

  b In the scales in the picture, above, three textbooks are balanced by four kilogram-bags of sugar. Estimate the weight of one textbook.

Scales and drawing

Scale drawing

- Scale drawings are used to give an **accurate representation** of a real object.

This scale drawing shows the plan of a bedroom.

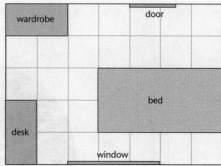

Scale: 1 cm represents 50 cm

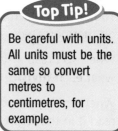

Top Tip!

Be careful with units. All units must be the same so convert metres to centimetres, for example.

- When you use a scale drawing make sure you know what the scale is.

- The actual measurement and the scaled measurement are connected by a **scale factor**.

- Scales are often given as **ratios**.

The wardrobe is 2 centimetres by 1 centimetre on the scale drawing.
The scale is 1 centimetre represents 50 centimetres, or 1:50.
So the wardrobe is 1 metre by 50 centimetres in the real bedroom.

F

Nets

- **Polyhedra** are solid shapes with **plane** (flat) sides.

- A **net** is a flat shape that can be folded into the **3-D shape**.

- Most **three-dimensional shapes**, particularly polyhedra, can be made from **nets**.

These are the nets of two common solids.

F-E

Questions

Grade F

1 Refer to the scale drawing of the bedroom, above.

 a How wide is the window?

 b What are the dimensions of the bed?

 c The desk is to be covered with a plastic sheet. The plastic costs £5 per square metre. How much will it cost to cover the desk?

Grade F

2 What two solids do the nets, above, represent?

3-D drawing

Isometric grids

- **Isometric grids** are usually drawn with solid lines or as a **dotty triangular grid**.

- An isometric grid can be used to show a **two-dimensional view** of a **three-dimensional object**.

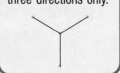

When questions using isometric grids are set in examinations the grid will be the right way round. Lines are always drawn in three directions only.

- When you use isometric grids it is important to ensure they are the **correct way round**, as in the diagram above.

Plans and elevations

- The **plan** of a shape is the view seen when looking directly down from above.

- The **front elevation** of a shape is the view seen when looking from the front.

- The **side elevation** of a shape is the view seen when looking from the side.

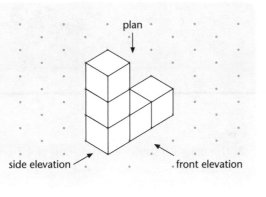

Questions

Grade E

1 Refer to the isometric drawing of the cuboid, above.

   a What are the lengths of the sides of the cuboid?

   b What is the volume of the cuboid?

Grade D

2 Refer to the isometric drawing, above. On centimetre-squared paper, draw:

   a the plan

   b the front elevation

   c the side elevation.

Congruency and tessellations

SHAPE, SPACE AND MEASURES

Congruent shapes

- When two shapes are **congruent** they have **exactly the same** dimensions.

- Congruent shapes can be **reflections** or **rotations** of each other.

These triangles are all congruent to each other.

G

Top Tip!

To check congruency, use tracing paper to copy one shape and place the tracing paper over the other shapes to see if they are exactly the same. You may have to turn the tracing paper over.

Tessellations

- When a shape **tessellates** it fits together so that there are **no overlaps** and **no gaps**.

- A **tessellation** is the pattern that is formed when the shapes are fitted together.

These are all tessellations.

E

Questions

Grade G

1 Are the shapes in each pair congruent?

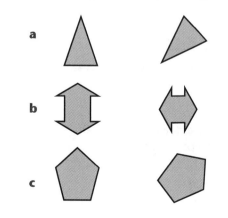

a

b

c

Grade E

2 The diagram shows a tessellation of a 1 cm by 2 cm rectangle.

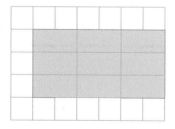

On centimetre-squared paper, draw a different tessellation of a 1 cm by 2 cm rectangle. Draw at least six rectangles to show your tessellation clearly.

Transformations

Transformations

- When a shape is **moved**, **rotated**, **reflected** or **enlarged**, this is a **transformation**. The original shape is called the **object**, and the transformed shape is called the **image**.

- The four transfomations used in GCSE exams are **translation**, **reflection**, **rotation** and **enlargement**.

- Translated, reflected or rotated shapes are **congruent** to the original.

- Shapes that are enlarged are **similar** to each other.

- Sometimes the transformation is said to **map** the object to the image.

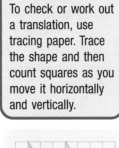

Translations

- When a shape is **translated** it is moved without altering its orientation; it is *not* rotated or reflected.

- A translation is described by **a vector**, for example, $\binom{-4}{5}$.

- The **top number** in the vector is the movement in the x-direction. Positive values move to the right. Negative values move to the left.

- The **second** or **bottom number** in the vector is the movement in the y-direction, or vertically. Positive values move upwards. Negative values move downwards.

  Triangle C is translated from triangle A by the vector $\binom{0}{4}$.

  Triangle B is translated to triangle D by the vector $\binom{-2}{4}$.

Top Tip!

To check or work out a translation, use tracing paper. Trace the shape and then count squares as you move it horizontally and vertically.

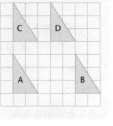

Reflections

- When a shape is **reflected** it becomes a **mirror image** of itself.

- A **reflection** is described in terms of a mirror line.

- **Equivalent points** on either side are the **same distance** from the mirror line and the **line joining them** crosses the **mirror line** at **right angles**.

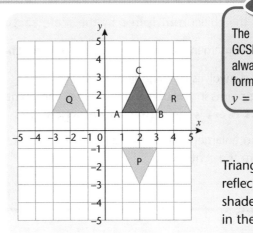

Top Tip!

The mirror lines in GCSE questions will always be of the form $y = a$, $x = b$, $y = x$, $y = -x$.

Triangle P is a reflection of the shaded triangle in the x-axis.

Questions

Grade D

1 Refer to the middle diagram, above.
  What vector translates:
  a triangle A to **i** triangle B **ii** triangle D
  b triangle C to **i** triangle B **ii** triangle A?

Grade D

2 Refer to the bottom diagram, above.
  What is the mirror line for the reflection:
  a that takes the shaded triangle to triangle Q
  b that takes the shaded triangle to triangle R?

Remember: You must revise all content from Grade G to the level that you are currently working at.

Rotations

D–E

- When a shape is **rotated** it is turned about a centre, called the **centre of rotation**.

- The rotation will be in a **clockwise** or **anticlockwise** direction.

- The rotation can be described by an **angle**, such as 90°, or a **fraction of a turn**, such as 'a half-turn'.

Triangle A is a rotation of the shaded triangle through 90° in a clockwise direction about the centre (0, 1).

- A rotation of 90° clockwise is the same as a rotation of 270° anticlockwise. Only a half-turn does not need to have a direction specified.

Top Tip!

To check or work out a rotation, use tracing paper. Use a pencil point on the centre and rotate the tracing paper in the appropriate direction, through the given angle.

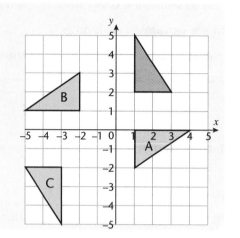

Enlargements

D–C

Top Tip!

Scale factors can also be fractions. In this case the enlargement makes things smaller!

- When a shape is **enlarged** it changes its size to become a shape that is **similar** to the first shape.

- An **enlargement** is described by a **centre** and a **scale factor**.

- The length of the sides of the image will be the length of the sides of the object **multiplied** by the **scale factor**.

- The centre of enlargement can be found by the **ray method**.

- The ray method involves drawing lines through corresponding points in the object and image. The point where the **rays meet** is the **centre of enlargement**.

Triangle Q is an enlargement of triangle P with a scale factor of 3 from the centre O.

Questions

Grade D

1 Refer to the top diagram, above. What is the rotation that takes:

  a the shaded triangle to triangle B

  b the shaded triangle to triangle C?

Grade D

2 Refer to the bottom diagram, above.

  a i Write down the scale factor of the enlargement that takes triangle P to triangle R.

  ii Mark the centre of the enlargement that takes triangle P to triangle R.

  b i Write down the scale factor of the enlargement that takes triangle Q to triangle R.

  ii Mark the centre of the enlargement that takes triangle Q to triangle R.

Constructions

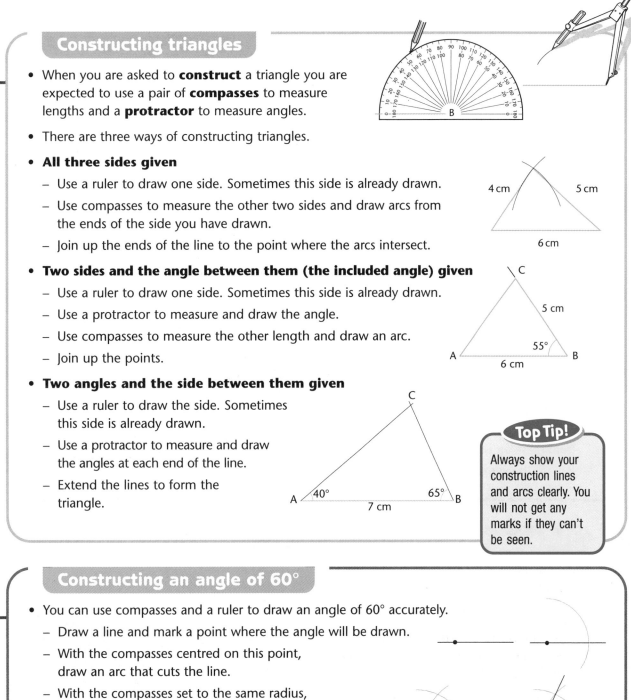

D

Constructing triangles

- When you are asked to **construct** a triangle you are expected to use a pair of **compasses** to measure lengths and a **protractor** to measure angles.

- There are three ways of constructing triangles.

- **All three sides given**
  - Use a ruler to draw one side. Sometimes this side is already drawn.
  - Use compasses to measure the other two sides and draw arcs from the ends of the side you have drawn.
  - Join up the ends of the line to the point where the arcs intersect.

- **Two sides and the angle between them (the included angle) given**
  - Use a ruler to draw one side. Sometimes this side is already drawn.
  - Use a protractor to measure and draw the angle.
  - Use compasses to measure the other length and draw an arc.
  - Join up the points.

- **Two angles and the side between them given**
  - Use a ruler to draw the side. Sometimes this side is already drawn.
  - Use a protractor to measure and draw the angles at each end of the line.
  - Extend the lines to form the triangle.

Top Tip!

Always show your construction lines and arcs clearly. You will not get any marks if they can't be seen.

C

Constructing an angle of 60°

- You can use compasses and a ruler to draw an angle of 60° accurately.
  - Draw a line and mark a point where the angle will be drawn.
  - With the compasses centred on this point, draw an arc that cuts the line.
  - With the compasses set to the same radius, and centred on the point where the first arc cut the line, draw another arc to cut the first arc.
  - Join the original point to the point where the arcs cross.

Questions

(Grade D)

1 Use a ruler, compasses and protractor for these questions.

  a Draw accurately the triangle with sides of 4 cm, 5 cm and 6 cm, shown above.

  b Draw accurately the triangle with sides of 6 cm and 5 cm and an included angle (between them) of 55°, shown above.

  c Draw accurately the triangle with a side of 7 cm, and angles of 40° and 65°, shown above.

(Grade C)

2 Follow the steps shown, above, and construct an angle of 60°.
   Use a protractor to check the accuracy of your drawing.

The perpendicular bisector

C

- To **bisect** means to *divide in half*. **Perpendicular** means *at right angles*.
- A **perpendicular bisector** divides a line in two, and is at right angles to it.
  - Start off with a line or two points. (Usually, these are given.)
  - Open the compasses to a radius about three-quarters of the length of the line, or the distance between the points.
  - With the compasses centred on each end of the line (or each point) in turn, draw arcs on either side of the line.
  - Join up the points where the arcs cross.

The angle bisector

C

- An **angle bisector** divides an angle into two smaller, equal angles.
  - Start off with an angle. (Usually, this is given.)
  - Open the compasses to any radius shorter than the arms of the angle. With the compasses centred on the **vertex** of the angle (the point where the arms meet, draw an arc on each arm of the angle).
  - Now, with the compasses still set to the same radius, draw arcs from these arcs so they intersect.
  - Join up the vertex of the angle and the point where the arcs cross.

The perpendicular at a point on a line

C

- The perpendicular from a point on a line is a line at right angles to the line, passing through the point.

  - Start off with a point on a line. (Usually, this is given.)

  - Open the compasses to a radius of about 3 cm and draw arcs on the line, centred on the point and on either side of it.

  - Now increase the radius of the compasses to about 5 cm and draw arcs centred on the two points on either side, so that they intersect.

  - Join up the point on the line and the point where the arcs cross.

Questions

(Grade C)

1 Draw a line 6 cm long. Following the steps above, draw the perpendicular bisector of the line. Check that each side is 3 cm long and that the angle is 90°.

(Grade C)

2 Draw an angle of 70°. Following the steps above,

draw the angle bisector of the line. Check that each half angle is 35°.

(Grade C)

3 Draw a line and mark a point on it. Following the steps above, draw the perpendicular from the point to the line. Check that the angle is 90°.

Constructions and loci

C

The perpendicular from a point to a line

- The **perpendicular** is a line that is at right angles to the original line and passes through the point.
  - Start with a line and a point not on the line. (Usually, these are given.)
  - Open the compasses to a radius about 3 cm more than the distance from the point to the line. With the compasses centred on the point, draw arcs on the line on either side of the point.
  - Centring the compasses on the points where these arcs cut the line, draw arcs on the other side of the line so they intersect.
  - Join the original point and the point where the arcs cross.

C

Loci

- A **locus** (plural **loci**) is the path followed by a point according to a rule.

  A point that moves so that it is always 4 cm from a fixed point is a circle of radius 4 cm.

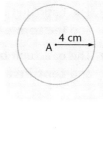

  A point that moves so that it is always the same distance from two fixed points is the perpendicular bisector of the two points.

A
•

B

C

Practical problems

- **Loci** can be used to solve real-life problems.

  A horse is tethered to a rope 10 m long in a large, flat field. What is the area the horse can graze?

  The horse will be able to graze anywhere within a circle of radius 10 m.

- In reality the horse may not be able to graze an exact circle but the situation is **modelled** by the mathematics.

Questions

(Grade C)

1 Following the steps, above, draw the perpendicular from a point to a line.
Check that the angle is 90°.

(Grade C)

2 A radar station in Edinburgh has a range of 200 miles.
A radar station in London has a range of 250 miles.
London and Edinburgh are 400 miles apart.
Sketch the area that the radar stations can cover.
Use a scale of 1 cm = 200 miles.

•Edinburgh

•London

Units

Systems of measurement

- The two main systems of measurement used in Britain are the **imperial system** and the **metric system**.

- The **imperial system** is based on measurements introduced many years ago. 12 inches = 1 foot, 16 ounces = 1 pound, 14 pounds = 1 stone.

- The **metric system** is used in Europe and most of the rest of the world. It is the standard system used in science and is based on the decimal system.

Top Tip!
Strictly speaking, what we refer to as *weight* should be called mass but the term *weight* is used in Foundation exams as it is in everyday use.

G-F

The metric system

- The basic unit of length is the **metre** (m). Other units are **millimetres** (mm), **centimetres** (cm) and **kilometres** (km).
  10 mm = 1 cm, 100 cm = 1 m, 1000 m = 1 km

- The basic unit of weight is the **kilogram** (kg). Other units are the **gram** (g) and the **tonne** (T).   1000 g = 1 kg, 1000 kg = 1 T

- The basic unit of capacity is the **litre** (l). Other units are **millilitres** (ml) and **centilitres** (cl).   1000 ml = 1 litre, 10 ml = 1 cl, 100 cl = 1 litre

Top Tip!
You should know the connection between the metric units.

G

The imperial system

- The basic unit of length is the **foot** (ft). Other units are **yards** (yd), **inches** (in) and **miles** (m). 12 in = 1 ft, 3 ft = 1 yd, 1760 yd = 1 m

- The basic unit of weight is the **pound** (lb). Other units are the **ounce** (oz), the **stone** (st) and the **ton** (ton).   16 oz = 1 lb, 14 lb = 1 st, 2240 lb = 1 ton

- The basic unit of capacity is the **pint** (pt). Other units are **gallons** (gall) and **quarts** (qt).   2 pt = 1 qt, 8 pt = 1 gall

Top Tip!
You do not need to know these conversions. If you need them in a question they will be given.

G

Conversions factors

- Because many imperial units, such as **miles** and **pounds**, are still in common use you need to be able to convert between them. To do this we use **conversion factors**.

- There are five conversion factors that you need to know:
  2.2 pounds ≈ 1 kilogram   1 foot ≈ 30 centimetres
  5 miles ≈ 8 kilometres   1.75 pints ≈ 1 litre   1 gallon ≈ 4.5 litres

Top Tip!
You need to know these five conversions. If you need any others in a question they will be given. However, another useful one to learn is 1 inch ≈ 2.5 centimetres. ≈ means 'approximately equal to'.

G

Questions

1 **a** Convert 120 centimetres to metres.
  b Convert 3500 grams to kilograms.
  c Convert 230 centilitres into litres.
  d Convert 45 millimetres into centimetres.

2 Use a calculator to work out approximately:
  a how many pounds are equivalent to 25 kg
  b how many litres are equivalent to 8 gallons
  c how many cm are equivalent to 1 yard
  d how many miles are equivalent to 120 km.

Surface area and volume of 3-D shapes

Units of length, area and volume

- The basic unit of length is the **metre** (m). Other units are **centimetres** (cm) and **millimetres** (mm).

  10 mm = 1 cm, 100 cm = 1 m

- The basic unit of area is the **square metre** (m^2). Other units are **square centimetres** (cm^2) and **square millimetres** (mm^2).

  100 mm^2 = 1 cm^2, 10 000 cm^2 = 1 m^2

- The basic unit of volume is the **cubic metre** (m^3). Other units are **cubic centimetres** (cm^3) and **cubic millimetres** (mm^3).

  1000 mm^3 = 1 cm^3, 1 000 000 cm^3 = 1 m^3

- You need to be able to relate the areas and volumes of similar shapes.

> **Top Tip!**
>
> When converting between units of volume, such as cubic centimetres and cubic metres, a common mistake is just to divide by 100 instead of 1 000 000 (100^3). Learn the connections.

These cubes are all similar.

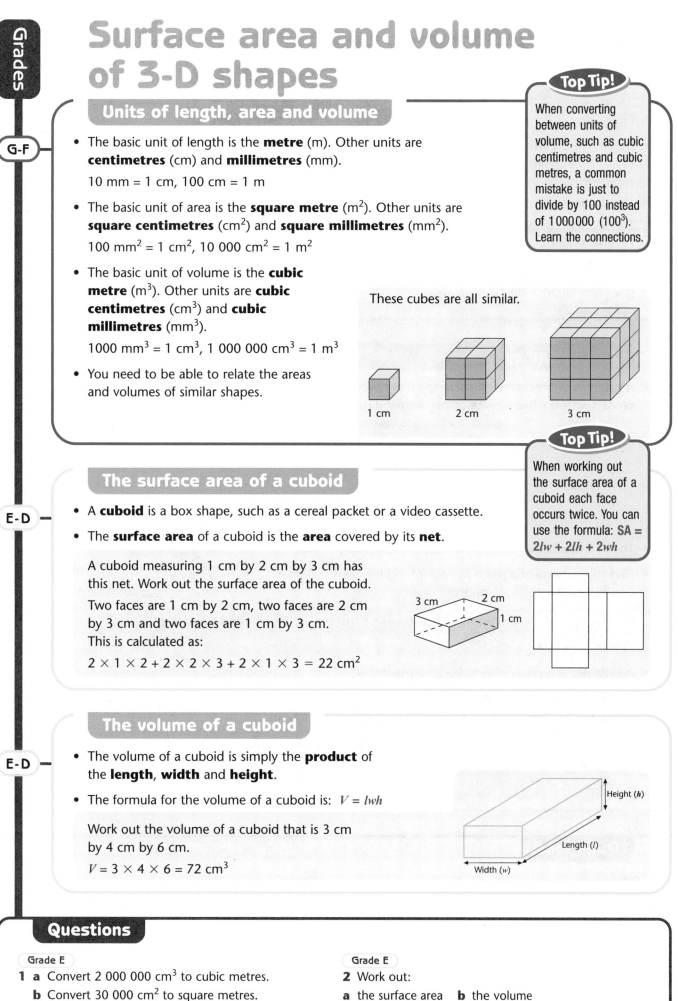

1 cm   2 cm   3 cm

The surface area of a cuboid

- A **cuboid** is a box shape, such as a cereal packet or a video cassette.

- The **surface area** of a cuboid is the **area** covered by its **net**.

  A cuboid measuring 1 cm by 2 cm by 3 cm has this net. Work out the surface area of the cuboid.

  Two faces are 1 cm by 2 cm, two faces are 2 cm by 3 cm and two faces are 1 cm by 3 cm. This is calculated as:

  $2 \times 1 \times 2 + 2 \times 2 \times 3 + 2 \times 1 \times 3 = 22$ cm^2

> **Top Tip!**
>
> When working out the surface area of a cuboid each face occurs twice. You can use the formula: **SA = 2lw + 2lh + 2wh**

3 cm   2 cm   1 cm

The volume of a cuboid

- The volume of a cuboid is simply the **product** of the **length**, **width** and **height**.

- The formula for the volume of a cuboid is: $V = lwh$

  Work out the volume of a cuboid that is 3 cm by 4 cm by 6 cm.

  $V = 3 \times 4 \times 6 = 72$ cm^3

Height (*h*)   Length (*l*)   Width (*w*)

Questions

Grade E

1 a Convert 2 000 000 cm^3 to cubic metres.
  b Convert 30 000 cm^2 to square metres.
  c Convert 5 m^3 to cubic centimetres.
  d Convert 4 cm^2 to square millimetres

Grade E

2 Work out:

  a the surface area b the volume
  of a cuboid above with sides of 2 cm, 3 cm and 4 cm.

Density and prisms

Top Tip!

Although there is a difference between mass and weight, in GCSE Foundation maths they are assumed to have the same meaning.

C

Density

- **Density** is the **mass** of a substance per unit **volume** and is usually expressed in **grams per cubic centimetre** (g/cm³) or kilograms per cubic metre (kg/m³).
- The relationship between density, mass and volume is: $\mathbf{density} = \dfrac{\mathbf{mass}}{\mathbf{volume}}$
- This triangle can be used to find mass and volume in terms of other quantities.

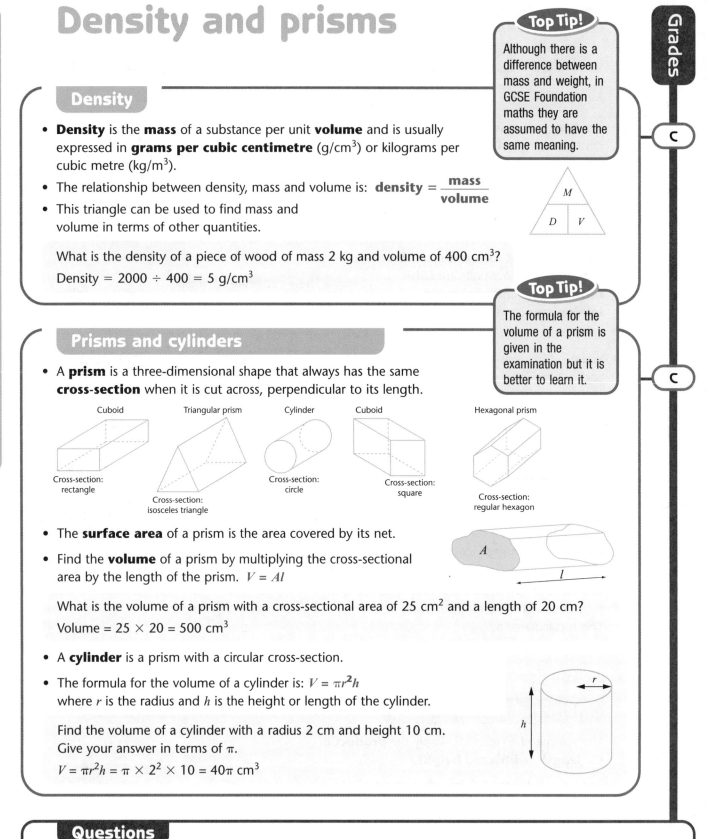

What is the density of a piece of wood of mass 2 kg and volume of 400 cm³?

Density = 2000 ÷ 400 = 5 g/cm³

Top Tip!

The formula for the volume of a prism is given in the examination but it is better to learn it.

C

Prisms and cylinders

- A **prism** is a three-dimensional shape that always has the same **cross-section** when it is cut across, perpendicular to its length.

Cuboid   Triangular prism   Cylinder   Cuboid   Hexagonal prism

Cross-section: rectangle

Cross-section: isosceles triangle

Cross-section: circle

Cross-section: square

Cross-section: regular hexagon

- The **surface area** of a prism is the area covered by its net.
- Find the **volume** of a prism by multiplying the cross-sectional area by the length of the prism. $V = Al$

What is the volume of a prism with a cross-sectional area of 25 cm² and a length of 20 cm?

Volume = 25 × 20 = 500 cm³

- A **cylinder** is a prism with a circular cross-section.
- The formula for the volume of a cylinder is: $V = \pi r^2 h$ where r is the radius and h is the height or length of the cylinder.

Find the volume of a cylinder with a radius 2 cm and height 10 cm. Give your answer in terms of π.

$V = \pi r^2 h = \pi \times 2^2 \times 10 = 40\pi$ cm³

Questions

Grade C

1 **a** What is the density of a brick with a mass of 2500 grams and a volume of 400 cm³? State the units of your answer.

  b A piece of metal has a density of 9 g/cm³. It has a volume of 3000 cm³. How much does it weigh? Give your answer in kilograms.

  c A piece of wood has an average density of 52 kg/m³. It weighs 15.6 kilograms. What is its volume? Give your answer in cubic metres.

Grade C

2 **a** A cuboid has dimensions 4 cm by 6 cm by 10 cm.

  i What is the cross-sectional area?

  ii What is the volume?

  b A triangular prism has a cross-sectional area of 3.5 m² and a length of 12 m. What is its volume?

Grade C

3 What is the volume of a cylinder with a radius of 4 cm and a height of 10 cm? Give your answer in terms of π.

Pythagoras' theorem

Pythagoras' theorem

- **Pythagoras' theorem** connects the sides of a **right-angled triangle**.

- Pythagoras' theorem states that: 'In any right-angled triangle, the square of the hypotenuse is equal to the sum of the squares of the other two sides.'

- The **hypotenuse** is the **longest** side of the triangle, which is always **opposite** the right angle.

- Pythagoras' theorem is usually expressed as a formula: $c^2 = a^2 + b^2$

  where c is the length of the hypotenuse and a and b are the lengths of the other two sides.

- The formula can be rearranged to find one of the other sides: $a^2 = c^2 - b^2$

> **Top Tip!**
> To find the actual value of a side, don't forget to take the square root.

Finding lengths of sides

- If you know the lengths of two sides of a right-angled triangle then you can always use Pythagoras' theorem to find the length of the third side.

  Find the length of the hypotenuse of this triangle.

  Using Pythagoras' theorem:
  $x^2 = 9^2 + 6.2^2 = 81 + 38.44$
  $\quad = 119.44$
  $x = \sqrt{119.44} = 10.9$ cm

  Find the length of the side marked y in this triangle.

  Using Pythagoras' theorem:
  $y^2 = 16^2 - 12^2 = 256 - 144$
  $\quad = 112$
  $y = \sqrt{112} = 10.6$

> **Top Tip!**
> Notice that when you are finding the **hypotenuse you add** the squares of the other two sides and when you are finding one of the **short sides, you subtract** the squares of the other two sides.

Real-life problems

- Pythagoras' theorem can be used to solve practical problems.

  A ladder of length 5 m is placed with the foot 2.2 m from the base of a wall. How high up the wall does the ladder reach?

  Using Pythagoras' theorem:
  $x^2 = 5^2 - 2.2^2$
  $\quad = 25 - 4.84$
  $\quad = 20.16$
  $x = \sqrt{20.16} = 4.5$ m

> **Top Tip!**
> In GCSE exams you should be given a diagram with a real-life Pythagoras' theorem problem. If not, draw one.

Questions

Grade C

1 A ladder is placed 1.5 m from the base of the wall and reaches 3.6 m up the wall. How long is the ladder?

Shape, space and measures checklist

I can...

- [ ] find the perimeter of a 2-D shape
- [ ] find the area of a 2-D shape by counting squares
- [ ] draw lines of symmetry on basic 2-D shapes
- [ ] use the basic terminology associated with circles
- [ ] draw circles with a given radius
- [ ] recognise the net of a simple shape
- [ ] name basic 3-D solids
- [ ] recognise congruent shapes
- [ ] find the volume of a 3-D shape by counting squares

You are working at (Grade G) level.

- [ ] find the area of a rectangle, using the formula $A = lw$
- [ ] find the order of rotational symmetry for basic 2-D shapes
- [ ] measure and draw angles accurately
- [ ] use the facts that the angles on a straight line add up to 180° and the angles around a point add up to 360°
- [ ] draw and measure lines accurately
- [ ] draw the net of a simple 3-D shape
- [ ] read scales with a variety of divisions
- [ ] find the surface area of a 2-D shape by counting squares

You are working at (Grade F) level.

- [ ] find the area of a triangle using the formula $A = \frac{1}{2}bh$
- [ ] draw lines of symmetry on more complex 2-D shapes
- [ ] find the order of rotational symmetry for more complex 2-D shapes
- [ ] measure and draw bearings
- [ ] use the facts that the angles in a triangle add up to 180° and the angles in a quadrilateral add up to 360°
- [ ] find the exterior angle of a triangle and a quadrilateral
- [ ] recognise and find opposite angles
- [ ] draw simple shapes on an isometric grid
- [ ] tessellate a simple 2-D shape
- [ ] reflect a shape in the x- and y-axes
- [ ] convert from one metric unit to another
- [ ] convert from one imperial unit to another, given the conversion factor
- [ ] use the formula $V = lwh$ to find the volume of a cuboid
- [ ] find the surface area of a cuboid

You are working at (Grade E) level.

- [ ] find the area of a parallelogram, using the formula $A = bh$
- [ ] find the area of a trapezium, using the formula $\frac{1}{2}(a + b)h$
- [ ] find the area of a compound shape
- [ ] work out the formula for the perimeter, area or volume of simple shapes

- [ ] identify the planes of symmetry for 3-D shapes
- [ ] recognise and find alternate angles in parallel lines and a transversal
- [ ] recognise and find corresponding angles in parallel lines and a transversal
- [ ] recognise and find interior angles in parallel lines and a transversal
- [ ] use and recognise the properties of quadrilaterals
- [ ] find the exterior and interior angles of regular polygons
- [ ] understand the words 'sector' and 'segment' when used with circles
- [ ] calculate the circumference of a circle, giving the answer in terms of π if necessary
- [ ] calculate the area of a circle, giving the answer in terms of π if necessary
- [ ] recognise plan and elevation from isometric and other 3-D drawings
- [ ] translate a 2-D shape
- [ ] reflect a 2-D shape in lines of the form $y = a$, $x = b$
- [ ] rotate a 2-D shape about the origin
- [ ] enlarge a 2-D shape by a whole-number scale factor about the origin
- [ ] construct diagrams accurately, using compasses, a protractor and a straight edge
- [ ] use the appropriate conversion factors to change imperial units to metric units and vice versa

You are working at (Grade D) level.

- [ ] work out the formula for the perimeter, area or volume of complex shapes
- [ ] work out whether an expression or formula represents a length, an area or a volume
- [ ] relate the exterior and interior angles in regular polygons to the number of sides
- [ ] find the area and perimeter of semicircles
- [ ] translate a 2-D shape, using a vector
- [ ] reflect a 2-D shape in the lines $y = x$, $y = -x$
- [ ] rotate a 2-D shape about any point
- [ ] enlarge a 2-D shape by a fractional scale factor
- [ ] enlarge a 2-D shape about any centre
- [ ] construct perpendicular and angle bisectors
- [ ] construct an angle of 60°
- [ ] construct the perpendicular to a line from a point on the line and a point to a line
- [ ] draw simple loci
- [ ] work out the surface area and volume of a prism
- [ ] work out the volume of a cylinder, using the formula $V = \pi r^2 h$
- [ ] find the density of a 3-D shape
- [ ] find the hypotenuse of a right-angled triangle, using Pythagoras' theorem
- [ ] find the short side of a right-angled triangle, using Pythagoras' theorem
- [ ] use Pythagoras' theorem to solve real-life problems

You are working at (Grade C) level.

Basic algebra

The language of algebra

- **Algebra** is the use of **letters** instead of numbers to write rules, formulae and expressions.

- Algebra follows the same rules as arithmetic and uses the **same symbols** but there are some special rules in algebra.

  - 3 more than x is written as $x + 3$ or $3 + x$
  - 5 less than x is written as $x - 5$
  - p minus q is written as $p - q$
  - 8 times y is written as $8 \times y$ or $y \times 8$ or $8y$
  - b divided by 4 is written as $b \div 4$ or $\dfrac{b}{4}$
  - $1 \times n$ is just written as n
  - t times t is written as $t \times t$ or t^2

F

Simplifying expressions

- **Simplifying** an expression means writing it in as neat a form as possible.

- Simplify expressions by **collecting like terms** or **multiplying expressions**.

E-D

Collecting like terms

- To **collect like terms** combine terms that are similar, such as number terms or x-terms.

  a, $4a$ and $9a$ are like terms, $\frac{1}{2}x^2$ and $12x^2$ are like terms, $2ab$, $-3ba$ and $7ba$ are like terms.

- The number in front of the term is called the **coefficient**.

- **Group** all like terms together then **add** the **coefficients** in each group.

  To simplify $4a + 5b - 8 + 7a - 3b - 6$ rewrite as $4a + 7a + 5b - 3b - 8 - 6 = 11a + 2b - 14$.

E

Multiplying expressions

- To **multiply expressions**, multiply the numbers and use the index laws to multiply the letters.

  To simplify $4a^2b \times 3a^3b^2$ rewrite as $4 \times 3 \times a^2 \times a^3 \times b \times b^2 = 12a^5b^3$.

D

Questions

Grade F

1 Write the following as algebraic expressions.
   a p less than r b 7 plus x
   c a times b d t divided by 2
   e n more than m

Grade E

2 Simplify the following by collecting like terms.
   a $x + 2x + 5x$ b $6x + 9 + 3x - 4$
   c $6w - 3k - 2w - 3k + 5w$ d $9x^2 + 6z - 8z - 7x^2$

Grade D

3 Simplify the following by multiplying the expressions.
   a $3 \times 4t$ b $4n^2 \times 3n$
   c $5mn \times 6m$ d $-3x^2y \times 4xy^3$

Expanding and factorising

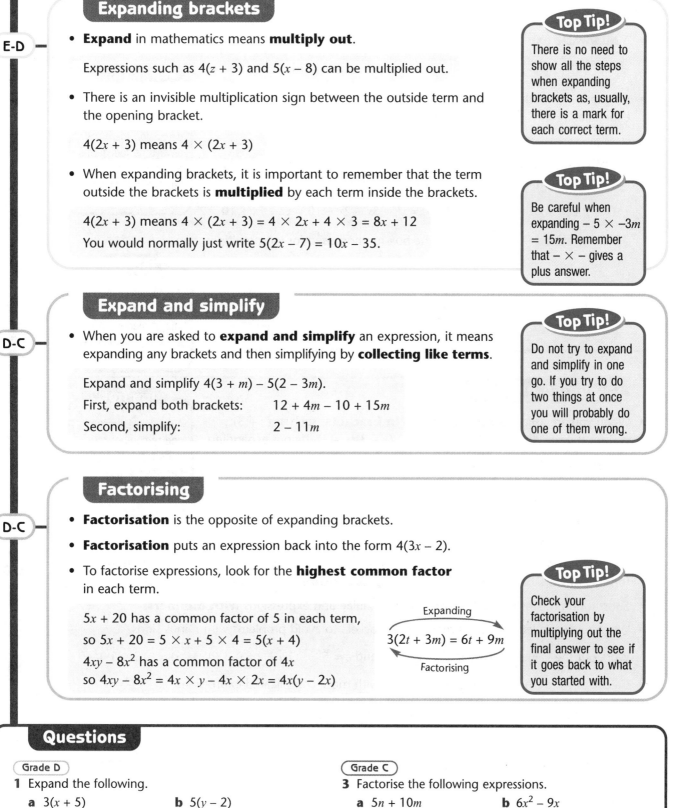

Expanding brackets

E-D

- **Expand** in mathematics means **multiply out**.

  Expressions such as $4(z + 3)$ and $5(x - 8)$ can be multiplied out.

- There is an invisible multiplication sign between the outside term and the opening bracket.

  $4(2x + 3)$ means $4 \times (2x + 3)$

- When expanding brackets, it is important to remember that the term outside the brackets is **multiplied** by each term inside the brackets.

  $4(2x + 3)$ means $4 \times (2x + 3) = 4 \times 2x + 4 \times 3 = 8x + 12$
  You would normally just write $5(2x - 7) = 10x - 35$.

Top Tip!
There is no need to show all the steps when expanding brackets as, usually, there is a mark for each correct term.

Top Tip!
Be careful when expanding $- 5 \times -3m = 15m$. Remember that $- \times -$ gives a plus answer.

Expand and simplify

D-C

- When you are asked to **expand and simplify** an expression, it means expanding any brackets and then simplifying by **collecting like terms**.

  Expand and simplify $4(3 + m) - 5(2 - 3m)$.
  First, expand both brackets: $12 + 4m - 10 + 15m$
  Second, simplify: $2 - 11m$

Top Tip!
Do not try to expand and simplify in one go. If you try to do two things at once you will probably do one of them wrong.

Factorising

D-C

- **Factorisation** is the opposite of expanding brackets.

- **Factorisation** puts an expression back into the form $4(3x - 2)$.

- To factorise expressions, look for the **highest common factor** in each term.

  $5x + 20$ has a common factor of 5 in each term,
  so $5x + 20 = 5 \times x + 5 \times 4 = 5(x + 4)$
  $4xy - 8x^2$ has a common factor of $4x$
  so $4xy - 8x^2 = 4x \times y - 4x \times 2x = 4x(y - 2x)$

  Expanding
  $3(2t + 3m) = 6t + 9m$
  Factorising

Top Tip!
Check your factorisation by multiplying out the final answer to see if it goes back to what you started with.

Questions

Grade D

1 Expand the following.

  a $3(x + 5)$ **b** $5(y - 2)$
  c $3(2x + y)$ **d** $n(n - 7)$
  e $5m(2m + 3)$ **f** $3p^2(2p - 3q)$

Grade C

2 Expand and simplify.

  a $2(x - 3) + 4(x + 3)$ **b** $5(m + 2) - 2(m - 6)$
  c $n(n + 1) + 3n(n - 2)$ **d** $6(x + 5) - 3(x + 1)$
  e $3x(2x - 3y) - 2x(x - y)$ **f** $8(x + 3y) + 2(x + 7y)$

Grade C

3 Factorise the following expressions.

  a $5n + 10m$ **b** $6x^2 - 9x$
  c $5mn + 6m$ **d** $4x^2y + 12xy^2$
  e $2xy + 6x^2$ **f** $2a^2b - 8ab + 6ab^2$

Quadratic expansion and substitution

Quadratic expansion

- A **quadratic expression** is one in which the highest power of any term is 2.

x^2, $4y^2 + 3y$, $6x^2 - 2x + 1$ and $4xy$ are quadratic expressions.

- An expression such as $(x + 2)(x - 3)$ can be **expanded** to give a quadratic expression. This is called **quadratic expansion**.

- Here are two methods for quadratic expansion.

  - **Splitting the brackets** The terms inside the first brackets are split and used to multiply the terms in the second brackets.

    $(x + 2)(x + 5) = x(x + 5) + 2(x + 5) = x^2 + 5x + 2x + 10 = x^2 + 7x + 10$

  - **Box method** This is similar to the box method used for long multiplication.

$(x - 4)(x - 2)$

×	x	$- 4$
x	x^2	$- 4x$
$- 2$	$- 2x$	$+ 8$

$= x^2 - 4x - 2x + 8$
$= x^2 - 6x + 8$

> **Top Tip!**
> When you multiply out a quadratic expansion there will always be four terms and two of these terms will combine together.

> **Top Tip!**
> Watch out for minus signs. This is where lots of marks are lost in exams. Remember: $-2 \times -4 = +8$.

Squaring brackets

- When you are asked to **square a term in brackets**, such as $(x + 3)^2$, you must write down the brackets twice, $(x + 3)(x + 3)$, before expanding.

$(x + 3)^2 = (x + 3)(x + 3) = x^2 + 3x + 3x + 9 = x^2 + 6x + 9$

> **Top Tip!**
> Always use brackets and remember the rules for dealing with negative numbers.

Substitution

- Formulae in mathematics are expressed algebraically. The area of a rectangle is $A = lb$. To use a formula to work out a value, such as an area, **substitute** numbers into the formula.

- Substitution means **replacing letters** in formulae and expressions **with numbers**.

- When replacing letters with numbers, use brackets to avoid problems with minus signs.

Work out the value of $ab + c$ if $a = -3$, $b = 4$ and $c = 5$. $ab + c = (-3)(4) + (5) = -12 + 5 = -7$

- Calculators have brackets keys that you use with more complicated expressions.

5 × (6 + 3) ÷ 3 = **15**

Questions

Grade C

1 Use whichever method you prefer to expand these.

  a $(x - 1)(x + 3)$ **b** $(m + 2)(m - 6)$

  c $(n + 1)(n - 2)$ **d** $(x + 5)(x + 1)$

  e $(x - 3)(x - 3)$ **f** $(x + 3)(x + 7)$

Grade C

2 Expand the following expressions.

  a $(x + 3)^2$ **b** $(x - 2)^2$

  c $(x - 1)(x + 1)$ **d** $(m + 2)(m - 2)$

Grade D

3 Using $x = 3$, $y = -4$ and $z = 5$, work out the value of:

  a $2x + 3y$ **b** $x^2 - yz$

  c $x(3y + 4z)$

Linear equations

Solving linear equations

C

- An **equation** is formed when an expression is put equal to a number or another expression.

- A **linear equation** is one that only involves one **variable**.

  $2x + 3 = 7$, $5x + 8 = 3x - 2$ are linear equations.

- **Solving** an equation means finding the value of the variable that makes it true.

  Solve $2x + 3 = 7$.

  The value of x that makes this true is 2 because $2 \times 2 + 3 = 7$

- The four ways to solve equations are shown below. These are all basically the same, but the most efficient is **rearrangement**.

Rearrangement

Solve $6x + 5 = 14$.

Move the 5 across the equals sign to give:

$6x = 14 - 5 = 9$

Divide both sides by 6 to give: $x = \frac{9}{6} \Rightarrow x = 1\frac{1}{2}$

Solve $\frac{y - 5}{3} = 6$.

Multiply both sides by 3 to give:

$y - 5 = 3 \times 6 = 18$

Move the 5 across the equals sign to give: $y = 18 + 5 = 23$

> **Top Tip!**
>
> This is called 'change sides, change signs', which means that plus becomes minus (and vice versa) and multiplication becomes division (and vice versa).

Inverse operations

Use the inverse operation method on simple equations such as $x + 7 = 9$ or $2x = 12$.
To solve the equation the opposite operation is applied to the right-hand side.

The inverse operation of +7 is –7 and of $2 \times x$ is 'divide by 2'.

$x + 7 = 9 \Rightarrow x = 9 - 7 = 2$
$2x = 12 \Rightarrow x = 12 \div 2 = 6$

Doing the same thing to both sides

Solve $3x - 5 = 16$.

Add 5 to both sides: $3x - 5 + 5 = 16 + 5$
$\qquad\qquad\qquad\qquad\quad 3x = 21$

Divide both sides by 3: $\quad 3x \div 3 = 21 \div 3$
$\qquad\qquad\qquad\qquad\qquad\quad x = 7$

Inverse flow diagram

Solve $3x - 4 = 11$.

Flow diagram: $\longrightarrow \boxed{\times 3} \longrightarrow \boxed{-4} \longrightarrow$

Inverse flow diagram: $\longleftarrow \boxed{\div 3} \longleftarrow \boxed{+4} \longleftarrow$

Put through: $\overset{5}{\longleftarrow} \boxed{\div 3} \overset{15}{\longleftarrow} \boxed{+4} \overset{11}{\longleftarrow}$

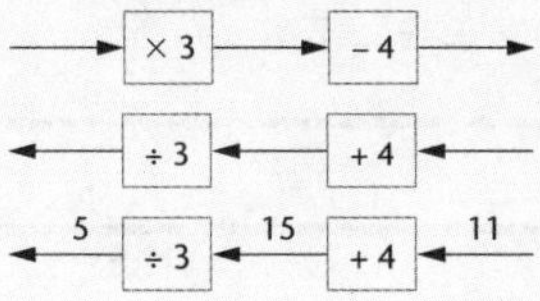

So the answer is 5.

> **Top Tip!**
>
> Always check that your answer works in the original equation.

Questions

1 Solve the following linear equations.

  a $3x - 5 = 4$ **b** $2m + 8 = 6$

  c $\frac{n}{3} = 4$ **d** $8x + 5 = 9$

  e $\frac{x - 3}{5} = 2$ **f** $5x - 3 = 7$

  g $\frac{y + 2}{5} = 3$ **h** $\frac{x}{7} - 3 = 1$

  i $8 - 2x = 7$

Remember: You must revise all content from Grade G to the level that you are currently working at.

Solving equations with brackets

D

- When an equation contains brackets you must first multiply out the brackets and then solve the equation in the normal way.

  Solve $4(x + 3) = 30$.

  Multiply out: $4x + 12 = 30$

  Subtract 12: $\qquad 4x = 18$

  Divide by 4: $\qquad x = 4\frac{1}{2}$

Equations with the variable on both sides of the equals sign

D

- When a letter appears on **both sides** of an equation use the 'change sides, change signs' rule.

- Use the rule to collect all the terms containing the **variable** on the **left-hand side** of the equals signs and all **number terms** on the **right-hand side**.

- When terms move across the equals sign, the signs change.

  Rearranging $5x - 3 = 2x + 12$ gives $5x - 2x = 12 + 3$.

- After the equation is rearranged, collect the terms together and solve the equation in the usual way.

  $5x - 3 = 2x + 12$ is rearranged to $5x - 2x = 12 + 3 \Rightarrow 3x = 15 \Rightarrow x = 5$

Top Tip!

Be careful when moving terms across the equals sign and remember to change the signs from plus to minus and vice versa.

Equations with brackets and the variable on both sides

C

- When an equation contains brackets and variables on **both sides**, always expand the brackets first.

  Expanding $4(x - 2) = 2(x + 3)$ gives $4x - 8 = 2x + 6$.

- After expanding the brackets, rearrange the equation, collect the terms and solve the equation in the usual way.

  $4(x - 2) = 2(x + 3)$ is expanded to $4x - 8 = 2x + 6 \Rightarrow 4x - 2x = 8 + 6$
  $$\Rightarrow 2x = 14$$
  $$\Rightarrow x = 7$$

Questions

Grade D

1 Solve the following by any method you choose.

  a $3(x - 5) = 12$ **b** $2(m - 3) = 6$

  c $3(x + 5) = 9$ **d** $2(x - 3) = 2$

  e $6(x - 1) = 9$ **f** $4(y + 2) = 2$

Grade D

2 Solve the following by any method you choose.

  a $3x - 2 = x + 12$ **b** $5y + 5 = 2y + 11$

  c $8x - 1 = 5x + 8$ **d** $7x - 3 = 2x - 8$

  e $6x - 2 = x - 1$ **f** $3y + 7 = y + 2$

Grade C

3 Solve the following by any method you choose.

  a $5(x - 3) = 2x + 12$ **b** $6(x + 1) = 2(x - 3)$

  c $3(x + 5) = 2(x + 10)$ **d** $7(x - 2) = 3(x + 4)$

D-C

Remember: You must revise all content from Grade G to the level that you are currently working at.

Setting up equations

- Linear equations can be used to **model** many practical problems.

The angles in a triangle are given as $2x$, $3x$ and $4x$.

What is the largest angle in the triangle?

The angles in a triangle add up to 180°, so:

$2x + 3x + 4x = 180°$

This simplifies to: $\quad 9x = 180°$

Divide by 9: $\qquad x = 20°$

So the largest angle is $4x = 4 \times 20 = 80°$.

Top Tip!

Use the letter x when setting up an equation unless you are given another letter to use.

Fred is 29 years older than his daughter, Freda. Together their ages add up to 47.

Using the letter x to represent Freda's age, write down an expression for Fred's age.

Set up an equation in x and solve it to find Freda's age.

Freda's age is x.

Fred's age is $x + 29$.

The sum of both their ages is $x + x + 29 = 47$.

Collecting terms:

$2x + 29 = 47$

Subtracting 29 from both sides:

$2x = 18$

Dividing by 2:

$x = 9$

Questions

Grade D

1 The diagram shows a rectangle.

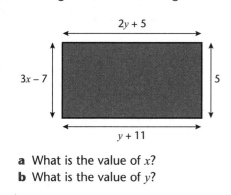

a What is the value of x?

b What is the value of y?

Grade D

2 A family has x bottles of milk delivered every day from Monday to Friday and seven bottles delivered on Saturday. There is no delivery on Sunday. In total they have 22 bottles delivered each week.

a Use the above information to set up an equation in x.

b Solve your equation to find the value of x.

Grade C

3 Asil thought of a number. He divided it by 2 then added 7. The result was 6 more than the number he first thought of.

a Use the information above to set up an equation, using x to represent the number Asil thought of.

b Solve your equation to find the value of x.

Trial and improvement

Trial and improvement

- Some equations **cannot be solved** simply by algebraic methods.

- A **numerical method** for solving these equations is called **trial and improvement**.

- After an initial guess (**the trial**), an answer is calculated and compared to the required answer.

- A better guess (**the improvement**) is then made.

- This process is **repeated** until the answer is within a given accuracy.

- Normally the first thing to do is find **two whole numbers** between which the answer lies.

- Then find **two decimal numbers**, **each with one decimal place**, between which the answer lies.

- Then test the **midway value** to see which of the two numbers is closer.

> **Top Tip!**
>
> In GCSE examinations one or two initial guesses are always given to give you a start.

Solve the equation $x^3 + 2x = 52$, giving your answer to one decimal place.

Start with a guess of $x = 4$: $4^3 + 2 \times 4 = 64 + 8 = 72$ This is too high.

Now try $x = 3$: $3^3 + 2 \times 3 = 27 + 6 = 33$ This is too low.

The answer must be between 3 and 4.

Now try $x = 3.5$: $3.5^3 + 2 \times 3.5 = 49.875$ This is close but too low.

Now try $x = 3.6$: $3.6^3 + 2 \times 3.6 = 53.856$ This is close but too high.

The answer must be between 3.5 and 3.6.

To see which is closer, try $x = 3.55$:

$$3.55^3 + 2 \times 3.55 = 51.838\ 875$$

which is just too low.

Hence, the answer, correct to one decimal place, is 3.6.

- Set out the working in a table.

Guess	$x^3 + 2x$	Comment
4	72	Too high
3	33	Too low
3.5	49.875	Too low
3.6	53.856	Too high
3.55	51.838875	Too low

> **Top Tip!**
>
> You must test the 'halfway' value between the one-decimal-place values.

Questions

Grade C

1 Use trial and improvement to find the solution to $x^3 - 2x = 100$.

Give your answer correct to one decimal place.

Grade C

2 Use trial and improvement to find the solution to $x^3 + x = 20$.

Give your answer correct to one decimal place.

Formulae

Grades

ALGEBRA

Formulae, identities, expressions and equations

F-C

- **Formulae** (singular *formula*) are important in mathematics as they express many of the rules we use in a concise manner.

  The area of a rectangle: $A = lb$

- **Identities** are certain formulae that are true, whatever values are used for the variables.

- The sign for an identity is a **'three-line' equals sign**, $\equiv$.

  $(x + a)^2 \equiv x^2 + 2ax + a^2$ is an identity.

- An **expression** is any arrangement of letters and numbers. The separate parts of expressions are called **terms**.

  $4a + 5b$ and $6x^2 - 2$ are expressions, $4a$, $5b$, $6x^2$ and -2 are terms.

- An **equation** is an expression that is put equal to another expression or a number.

- Equations can be **solved** to find the value of the variable that makes them true.

  $3x + 9 = 8$ and $4x + 7 = 2x - 1$ are equations.

Top Tip!

Make sure you know the difference between formulae, identities, expressions and equations. They could be tested in an examination.

Rearranging formulae

C

- The **subject** of a formula is the variable (letter) in the formula that stands on its own, usually on the left-hand side.

  x is the subject of each of these formula:

  $$x = 5t + 4 \qquad x = 4(2y - 7) \qquad x = \frac{1}{t}$$

- To change the subject of a formula, **rearrange** the formula to get the new variable on the left-hand side.

- To rearrange formulae, use the same rules as for solving equations.

- Unlike an equation, at each step the right-hand side is not a number but an expression.

  Make m the subject of $y = m + 3$.

  Subtract 3 from both sides: $y - 3 = m$

  Reverse the formula: $\qquad m = y - 3$

Top Tip!

Remember the rules, 'Change sides, change signs' and 'What you do to one side you must do to the other.'

Questions

Grade E

1 Which of the following are formulae, which are identities, which are expressions and which are equations?

  a $V = lwh$ **b** $3x + 8 = 7$ **c** $P = 2l + 2w$

  d $(x - 1)(x + 1) \equiv x^2 - 1$ **e** $4(x - 3) = 7$

  f $4x + 2y$ **g** $3(x + 1)^2$ **h** $c^2 \equiv a^2 + b^2$

Grade C

2 Rearrange each of the following formulae to make x the subject.

  a $T = 4x$ **b** $y = 2x + 3$ **c** $P = t + x$

  d $y = \dfrac{x}{5}$ **e** $A = mx$ **f** $S = 2\pi x$

Inequalities

Inequalities

- An **inequality** is an algebraic expression that uses the signs < (less than), > (greater than), ≤ (less than or equal to) and ≥ (greater than or equal to).

- The solution to an inequality is a **range of values**.

The expression $x < 2$ means that x can take any value less than 2, all the way to minus infinity. x can also be very close to 2, for example, 1.9999... but is never actually 2.

$x ≥ 3$ means that x can take the value of 3 itself or any value greater than 3, up to infinity.

Solving inequalities

- **Linear inequalities** can be solved by the same rules that you use to solve equations.

- The answer when a linear inequality is solved is an inequality such as $x > -1$.

Solve $2x + 3 < 11$.

$2x < 11 - 3 \Rightarrow 2x < 8 \Rightarrow x < 4$

Solve $\frac{x}{5} - 3 ≥ 4$.

$\frac{x}{5} ≥ 7 \Rightarrow x ≥ 35$

> **Top Tip!**
> Don't use equal signs when solving inequalities as this doesn't get any marks in an examination.

Inequalities on number lines

- The solution to a linear inequality can be shown on a **number line**.

- Use the convention that an open circle is a **strict inequality** and a filled-in circle is an **inclusive inequality**.

> **Top Tip!**
> Inequalities often ask for integer values. An integer is a positive or negative whole number, including zero.

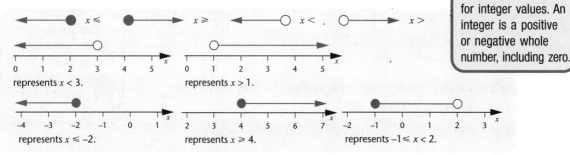

represents $x < 3$.

represents $x > 1$.

represents $x ≤ -2$.

represents $x ≥ 4$.

represents $-1 ≤ x < 2$.

Solve the inequality $2x + 3 < 11$ and show the solution on a number line.

The solution is $x < 4$, which is shown on this number line.

Questions

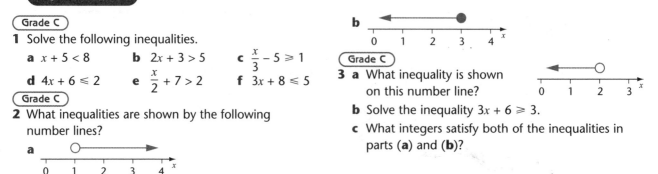

Grade C

1 Solve the following inequalities.

  a $x + 5 < 8$ **b** $2x + 3 > 5$ **c** $\frac{x}{3} - 5 ≥ 1$

  d $4x + 6 ≤ 2$ **e** $\frac{x}{2} + 7 > 2$ **f** $3x + 8 ≤ 5$

Grade C

2 What inequalities are shown by the following number lines?

  a

Grade C

3 a What inequality is shown on this number line?

  b Solve the inequality $3x + 6 ≥ 3$.

  c What integers satisfy both of the inequalities in parts (**a**) and (**b**)?

Graphs

Conversion graphs

• A conversion graph is a straight line graph that shows a relationship between two variables.

Graph 1 shows the relationship between litres and gallons.

Graph 2 shows the charge for units of electricity.

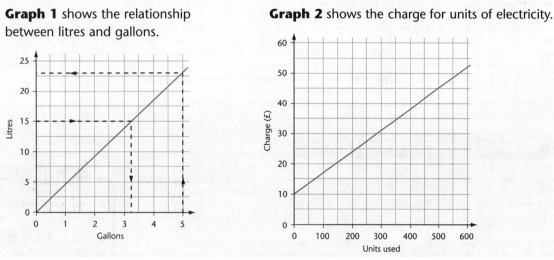

• Conversion graphs can be used to convert values from one unit to another.

In graph 1, 'litres and gallons', 15 litres is shown to be about $3\frac{1}{4}$ gallons and 5 gallons is shown to be about 23 litres.

Travel graphs

• A **travel graph** shows information about a journey.

• Travel graphs are also known as **distance–time** graphs.

• Travel graphs show the main features of a journey and use **average speeds**, which is why the lines in them are straight.

• In **reality**, vehicles do not travel at constant speeds.

• The average speed is given by:

$$\text{average speed} = \frac{\text{total distance travelled}}{\text{total time taken}}$$

• In a travel graph, the steeper the line, the **faster** the vehicle is travelling.

This graph shows a car journey from Barnsley to Nottingham and back.

Top Tip!

When asked for an average speed, answer in kilometres per hour (km/h) or miles per hour (mph).

Questions

Grade E

1 Refer to graph 2, above, showing the relationship between cost and units of electricity.

a How much will a customer who uses 500 units be charged?

b How many units will a customer who is charged £20 have used?

Grade D

2 Refer to the travel graph, above.

a After how many minutes was the car 16 kilometres from Barnsley?

b What happened between points D and E?

c During which part of the journey was the car travelling fastest?

d What was the average speed for the part of the journey between C and D?

Linear graphs

Negative coordinates

- You have already met **negative coordinates** and they often occur when drawing graphs.

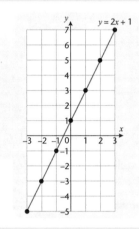

The coordinates of A are (1, 2), those of E are (−3, 2), H is (−3, −3) and J is (4, −2).

Drawing graphs from tables

The table shows values of the function $y = 2x + 1$ for values of x from −3 to +3.

Use the table to draw the graph of $y = 2x + 1$.

x	−3	−2	−1	0	1	2	3
y	−5	−3	−1	1	3	5	7

Top Tip!

Always label graphs. If you draw two graphs on the same set of axes and do not label them you will lose marks.

Drawing linear graphs

- You only need two points to draw a straight line.

- It is better to use three points, as the third point acts as a check.

Draw the graph of $y = 3x − 1$.

Pick values for x, such as $x = 3$ and work out the equivalent y-values.

$x = 3 \Rightarrow y = 3 \times 3 − 1 = 8$ giving (3, 8)

$x = 1 \Rightarrow y = 3 \times 1 − 1 = 2$ giving (1, 2)

$x = 0 \Rightarrow y = 3 \times 0 − 1 = −1$ giving (0, −1).

Top Tip!

In examinations you are always given a grid and told the range of x-values to use.

Top Tip!

Always use $x = 0$ as a point as it makes the calculation easy.

Questions

Grade E

1 Refer to the grid, above. Write down the coordinates of B, C, D, F, G and I.

Grade E

2 a Complete the table of values for the function $y = 2x − 3$ for values of x from −3 to + 3.

x	−3	−2	−1	0	1	2	3
y	−9	−7				1	3

b Draw a set of axes with x-values from −3 to +3 and y-values from −9 to +3. Use the table to draw the graph of $y = 2x − 3$.

Grade D

3 Draw a set of axes with x-values from −3 to +3 and y-values from −11 to +13.

Draw the graph of $y = 4x + 1$ for values of x from −3 to +3.

Remember: You must revise all content from Grade G to the level that you are currently working at.

F–D

Gradients

- The gradient of a line is a measure of how **steep** the line is.

- It is calculated by **dividing** the **vertical** distance between two points on the line by the **horizontal** distance between the same two points.

$$\text{gradient} = \frac{\text{vertical distance}}{\text{horizontal distance}}$$

- This is sometimes written as **gradient** $= \dfrac{y\text{-step}}{x\text{-step}}$

These lines have gradients as shown.

- Lines that slope down from left to right have **negative** gradients.

- The **right-angled triangles** drawn along grid lines are used to find gradients.

- To draw a line with a certain **gradient**, for every unit moved **horizontally**, move upwards (or downwards if the gradient is negative) by the number of units of the gradient.

Draw lines with gradients of $\frac{1}{4}$ and -2.

C

The gradient–intercept method

- The **gradient–intercept** method is the easiest and quickest method for drawing graphs.

- In the function $y = 2x + 3$, the **coefficient** of x (2) is the **gradient** and the **constant** term (+3) is the **intercept**.

- The intercept is the point where the line **crosses the y-axis**. (See the example below.)

C

Drawing a line with a given gradient

Draw the line $y = 2x + 3$.

Start by marking the intercept point (0, 3).

Next, move 1 unit across and 2 units up to show the gradient.

Repeat this a few more times.

Join up the points to get the required line.

Top Tip!
Draw graphs with a sharp pencil and use a ruler to make sure the lines are straight.

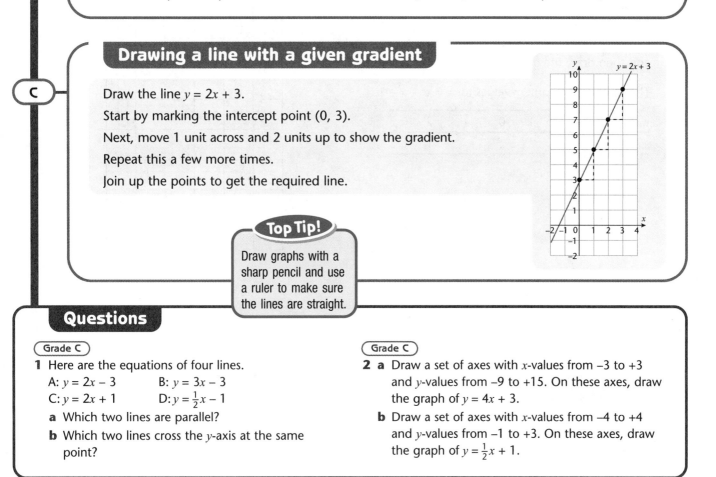

Questions

Grade C

1 Here are the equations of four lines.

A: $y = 2x - 3$ B: $y = 3x - 3$

C: $y = 2x + 1$ D: $y = \frac{1}{2}x - 1$

a Which two lines are parallel?

b Which two lines cross the y-axis at the same point?

Grade C

2 a Draw a set of axes with x-values from -3 to $+3$ and y-values from -9 to $+15$. On these axes, draw the graph of $y = 4x + 3$.

b Draw a set of axes with x-values from -4 to $+4$ and y-values from -1 to $+3$. On these axes, draw the graph of $y = \frac{1}{2}x + 1$.

Quadratic graphs

Drawing quadratic graphs

- A quadratic graph has an x^2 term in its equation.

  $y = x^2$, $y = x^2 + 2x + 3$ will give quadratic graphs.

- Quadratic graphs always have the same characteristic shape, which is called a **parabola**.

Top Tip!

Try to draw a smooth curve through all the points. Examiners prefer a good attempt at a curve, rather than points joined with a ruler.

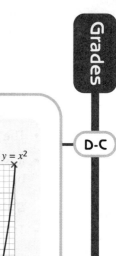

The graph of $y = x^2$

- Quadratic graphs are drawn from tables of values.

  This table shows the values of $y = x^2 + 2x - 3$ for values of x from -4 to 2.

x	-4	-3	-2	-1	0	1	2
y	5	0	-3	-4	-3	0	5

- Usually you will be asked to fill in some values in a table and then draw the graph.

  The table shows some y-values for $y = x^2 - x - 1$.

x	-3	-2	-1	0	1	2	3
y	11	5				1	5

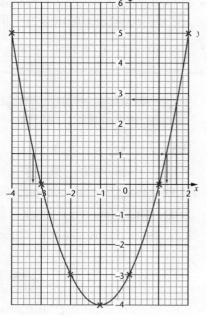

The graph of $y = x^2 + 2x - 3$

Top Tip!

As all parabolas have line symmetry, there will be some symmetry in the y-values in the table.

Questions

Grade C

1 a Complete the table of values above for $y = x^2 - x - 1$.

b Draw the graph of $y = x^2 - x - 1$.
Label the x-axis from 3 to +3 and the y-axis from -2 to 12.

Reading values from quadratic graphs

- Once a quadratic graph is drawn it can be used to solve various equations.

Use the graph of $y = x^2 - 2$ to find the x-values when $y = 5$.

Top Tip!

Always show the lines and arrows because, even if you make an error drawing the graph, you can still get marks for reading values from your graph.

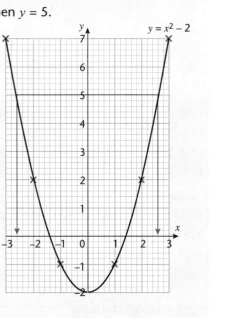

Draw the line $y = 5$ and draw down to the x-axis from the points where the line intercepts the curve.

The x-values are about +2.6 and –2.6.

Using graphs to solve quadratic equations

- Solving a quadratic equation means finding the x-values that make it true.

- To solve a quadratic equation from its graph, read the values where the curve crosses the x-axis.

Solve the equation $x^2 - 3x - 4 = 0$.

Top Tip!

The graph is $y = x^2 - 3x - 4$ and the equation is $x^2 - 3x - 4 = 0$. The solution of the equation is found where the graph crosses the x-axis, where $y = 0$.

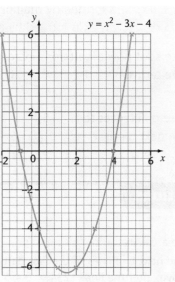

The graph crosses the x-axis at $x = -1$ and $x = 4$.

Questions

(Grade C)

1 a Use the graph of $y = x^2 - 2$ to solve the equation $x^2 - 2 = 0$.

b Use the graph of $y = x^2 - 3x - 4$ to find the x-values when $y = 2$.

Pattern

Patterns in number

- There are many curious and interesting patterns formed by numbers in mathematics.

$0 \times 9 + 1 = 1$
$1 \times 9 + 2 = 11$
$12 \times 9 + 3 = 111$
$123 \times 9 + 4 = 1111$
$1234 \times 9 + 5 = 11111$

$1 \times 8 + 1 = 9$
$12 \times 8 + 2 = 98$
$123 \times 8 + 3 = 987$
$1234 \times 8 + 4 = 9876$
$12345 \times 8 + 5 = 98765$

Top Tip!

When studying patterns formed by repeated calculations look for ways in which parts of the calculation continue in sequences.

$1 \times 3 \times 37 = 111$
$2 \times 3 \times 37 = 222$
$3 \times 3 \times 37 = 333$
$4 \times 3 \times 37 = 444$

$7 \times 7 = 49$
$67 \times 67 = 4489$
$667 \times 667 = 444889$
$6667 \times 6667 = 44448889$

- Use a calculator to check that these work. If you can spot the patterns, it is possible to write down the next term without using a calculator.

- Spotting patterns is an important part of mathematics. It helps in solving problems and making calculations.

Number sequences

- A **number sequence** is a series of numbers that build up according to some **rule**.

- A **linear sequence** is one in which terms increase by a **constant difference**.

  2, 8, 14, 20, 26, 32, ... is a linear sequence with an increase of +6 each time.

- Sometimes patterns build up according to more complicated rules.

  2, 4, 8, 16, 32, 64, ... is formed by doubling the previous term.
  2, 5, 11, 23, 47, ... is formed by multiplying the previous term by 2 and adding 1.
  1, 1, 2, 3, 5, 8, 13, 21, ... is formed by adding the two previous terms.

- One way to spot how a pattern is increasing is to look at the **differences** between consecutive terms.

  1   3   6   10   15   21
    2   3   4   5   6

- You can use differences to predict the next terms.

  15   21   28   36   45   55
    6   7   8   9   10

Questions

Grade E

1 Do not use a calculator for this question.
  a Write the next two lines of the first pattern, top left, above, after the line: $1234 \times 9 + 5 = 11111$
  b Write the next two lines of the second pattern, top right, above, after: $12345 \times 8 + 5 = 98765$
  c Write the next two lines of the third pattern, bottom left, above, after the line: $4 \times 3 \times 37 = 444$
  d Write the next two lines of the fourth pattern, bottom right, above, after the line: $6667 \times 6667 = 44448889$

Grade D

2 For each of the following patterns:
  i write down the next two terms
  ii describe how the pattern is building up.
  a 3, 9, 15, 21, 27, 33, ...
  b 3, 6, 12, 24, 48, ...
  c 100, 96, 92, 88, 84, 82, ...
  d 3, 5, 9, 15, 23, 33, 45, ...
  e 16, 13, 10, 7, 4, 1, ...

The nth term

C

The nth term of a sequence

- When using a number sequence, you may need to know the **50th** or **higher term**.

- This can be found by counting on but mistakes are likely, so it is easier to find the **nth term**.

- The nth term of is an **algebraic expression** that gives any term in the sequence.
  The nth term of a sequence is given as:

  $An + b$

  where A, the coefficient of n, is the difference between one term and the next term (**consecutive** terms) and b is the difference between the first term and A.

- To find any term **substitute** a number for n.

> The nth term of a sequence is found by the rule $3n + 1$.
> What are the first three terms and the 40th term?
> Take $n = 1$: $3 \times 1 + 1 = 4$
> Take $n = 2$: $3 \times 2 + 1 = 7$
> Take $n = 3$: $3 \times 3 + 1 = 10$
> Take $n = 40$: $3 \times 40 + 1 = 121$
> So the first three terms are 4, 7 and 10 and the 40th term is 121.

C

Finding the nth term

- The nth term of a linear sequence is always of the form $an \pm b$.

  $3n + 1$, $4n - 3$, $8n + 7$ are typical nth terms.

- To find the coefficient of n find the constant difference of the sequence.

  > The sequence 4, 7, 10, 13, 16, 19, ... has a constant difference of 3, so the nth term of the sequence will be given by $3n \pm b$.

- To find the value of b, work out what the difference is between the coefficient of n and the first term of the sequence.

  > The sequence 4, 7, 10, 13, 16, 19, ... has a first term of 4. The coefficient of n is 3. To get from 3 to 4, add 1, so the nth term is $3n + 1$.

  > What is the nth term of the sequence 4, 9, 14, 19, 24, 29, ...?
  > The constant difference is 5, and $5 - 1 = 4$, so the nth term is $5n - 1$.

Questions

Grade C

1 a The nth term of a sequence is $4n - 1$.

   i Write down the first three terms of the sequence.

   ii Write down the 100th term of the sequence.

b The nth term of a sequence is $\frac{1}{2}(n + 1)(n + 2)$.

   i Write down the first three terms of the sequence.

   ii Write down the 199th term of the sequence.

Grade C

2 Write down the nth term of each of these sequences.

   a 6, 11, 16, 21, 26, 31, ...

   b 3, 11, 19, 27, 35, ...

   c 9, 12, 15, 18, 21, 24, ...

Sequences

Special sequences

- There are many special sequences that you should be able to recognise.
  - The **even numbers**
    2, 4, 6, 8, 10, 12, 14, … The nth term is $2n$.
  - The **odd numbers**
    1, 3, 5, 7, 9, 11, 13, … The nth term is $2n - 1$.
  - The **square numbers**
    1, 4, 9, 16, 25, 36, 49, … The nth term is n^2.
  - The **triangular numbers**
    1, 3, 6, 10, 15, 21, 28, … The nth term is $\frac{1}{2}n(n + 1)$.
  - The **powers of 2**
    2, 4, 8, 16, 32, 64, 128, … The nth term is 2^n.
  - The **powers of 10**
    10, 100, 1000, 10 000, 100 000, … The nth term is 10^n.
  - The **prime numbers**
    2, 3, 5, 7, 11, 13, 17, 19, … There is no nth term as there is no pattern to the prime numbers.

> **Top Tip!**
> The only even prime number is 2

C

Finding the nth term from given patterns

- An important part of mathematics is to find **patterns in situations**.

- Once a **pattern** has been found the **nth term** can be used to describe the pattern.

The diagram shows a series of 'L' shapes.

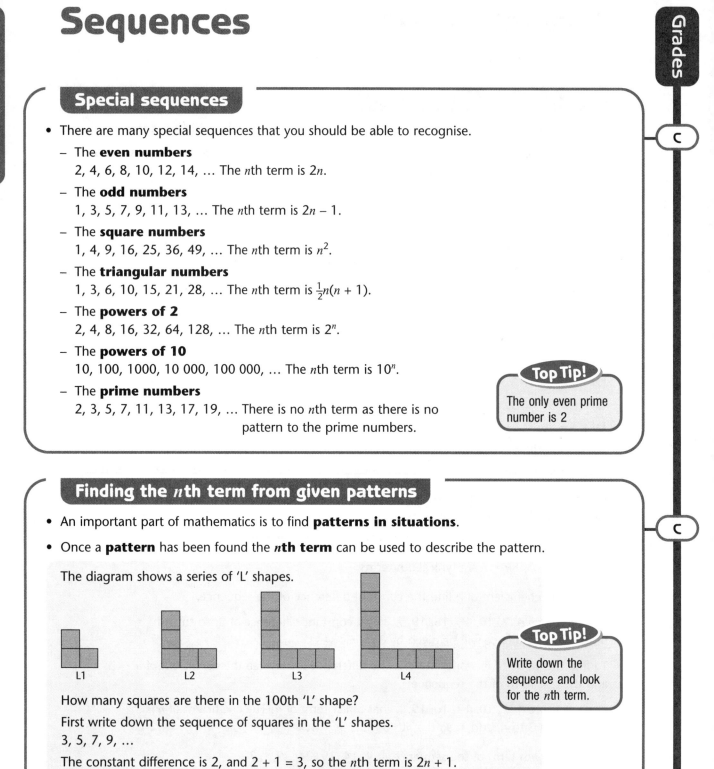

L1   L2   L3   L4

How many squares are there in the 100th 'L' shape?

First write down the sequence of squares in the 'L' shapes.

3, 5, 7, 9, …

The constant difference is 2, and $2 + 1 = 3$, so the nth term is $2n + 1$.

The 100 term will be $2 \times 100 + 1 = 201$, so there will be 201 squares in L100.

> **Top Tip!**
> Write down the sequence and look for the nth term.

C

Questions

Grade D

1 a What is the 100th even number?

b What is the 100th odd number?

c What is the 100th square number?

d Continue the sequence of triangular numbers for the next five terms.

e Continue the sequence of powers of 2 for the next five terms.

f Write down the next five prime numbers after 19.

Grade C

2 Matches are used to make pentagonal patterns.

1   2   3   4

a How many matches will be needed to make the 10th pattern?

b How many matches will be needed to make the nth pattern?

Algebra checklist

I can...

- [ ] use a formula expressed in words
- [ ] substitute numbers into expressions
- [ ] use letters to write a simple algebraic expression
- [ ] solve linear equations that require only one inverse operation to solve
- [ ] read values from a conversion graph
- [ ] plot coordinates in all four quadrants
- [ ] give the next value in a linear sequence
- [ ] describe how a linear sequence is building up

You are working at (Grade F) level.

- [ ] simplify an expression by collecting like terms
- [ ] simplify expressions by multiplying terms
- [ ] solve linear equations that require more than one inverse operation to solve
- [ ] read distances and times from a travel graph
- [ ] draw a linear graph from a table of values
- [ ] find any number term in a linear sequence
- [ ] recognise patterns in number calculations

You are working at (Grade E) level.

- [ ] use letters to write more complicated algebraic expressions
- [ ] expand expressions with brackets
- [ ] factorise simple expressions
- [ ] solve linear equations where the variable appears on both sides of the equals sign
- [ ] solve linear equations that require the expansion of a bracket
- [ ] set up and solve simple equations from real life situations
- [ ] find the average speed from a travel graph
- [ ] draw a linear graph without a table of values
- [ ] substitute numbers into an nth term rule
- [ ] understand how odd and even numbers interact in addition, subtraction and multiplication problems.

You are working at (Grade D) level.

- [ ] expand and simplify expressions involving brackets
- [ ] factorise expressions involving letters and numbers
- [ ] expand pairs of linear brackets to give a quadratic expression
- [ ] solve linear equations that have the variable on both sides and include brackets
- [ ] solve simple linear inequalities
- [ ] show inequalities on a number line
- [ ] solve equations using trial and improvement
- [ ] rearrange simple formulae
- [ ] use a table of values to draw a simple quadratic graph
- [ ] use a table of values to draw a more complex quadratic graph
- [ ] solve a quadratic equation from a graph
- [ ] give the nth term of a linear sequence
- [ ] give the nth term of a sequence of powers of 2 or 10.

You are working at (Grade C) level.

Answers

Page 4

1 a 28 and 18 **b** 1 tail and 1 head **c** 60

2 a 14 hours **b** ☺ ☺ ☺ ◖

Page 5

1 a Thursday and Saturday

 b 144

 c Friday

 d Wednesday, as most copies were sold.

2 a Town B, about 4 degrees hotter

 b Town B, summer is in December

 c September

 d No, the lines cross but they have no meaning, they just show trends.

Page 6

1 a 5 **b** Y

 c Because it is the smallest value.

2 a 7 **b** 8.5

3 a 64 **b** 35

Page 7

1 a 7 **b** 12

2 a Dan 76, Don 76 **b** Dan 100, Don 21

 c He could pick Dan as he can sometimes achieve high scores or he can pick Don because he is consistent and always achieves a reasonable score.

3 Asaf: median; Brian: mode; Clarrie: mean

Page 8

1 a 30 **b** 0 **c** 1 **d i** 45 **ii** 1.5

2 a $20 < x \leqslant 30$ **b i** 1350 **ii** 27

Page 9

1 a 16, 18, 19, 23, 26, 29, 30, 31, 32, 35, 38, 42, 42, 42, 57

1	8 6 9
2	3 6 9
3	0 1 2 5 8
4	2 2 2
5	7

 Key 4 | 2 = 42

 b 42 **c** It is an extreme value. **d** 31

 e 41 **f** 32

2 a 20 **b i** 13 **ii** 49 **iii** 36 **c** 35

 d 26.5 **e i** 560 **ii** 28

Page 10

1 a i Certain **ii** Evens **iii** Very likely

 iv impossible

b iii   **i**   **ii**

 0   0.5   1

2 a $\frac{1}{6}$ **b** $\frac{12}{52} = \frac{3}{13}$ **c** $\frac{1}{2}$

Page 11

1 $\frac{7}{10}$

2 a $\frac{4}{52} = \frac{1}{13}$ **b** $\frac{4}{52} = \frac{1}{13}$ **c** $\frac{8}{52} = \frac{2}{13}$

3 a Ali 0.25 Barry 0.22 Clarrie 0.19

 b Clarrie, most trials **c** 40

Page 12

1 a 36 **b i** $\frac{6}{36} = \frac{1}{6}$ **ii** $\frac{4}{36} = \frac{1}{9}$

 c i $\frac{4}{36} = \frac{1}{9}$ **ii** $\frac{6}{36} = \frac{1}{6}$ **iii** 7

Page 13

1 a 50 **b** 20 **c** 70

2 a i 5 **ii** 30 **iii** 13 **iv** $\frac{12}{30} = \frac{2}{5}$

 b i 200 **ii** 5% **iii** 40%

Page 14

1 a 120 **b** 33°

Page 15

1 a The scatter diagram shows positive correlation.

 b Students who can learn their tables can usually also learn spellings.

2 a As a car gets older, its value decreases.

 b There is no relationship between someone's wages and the distance they live from work.

3 73

Page 16

1 a Leading question, two questions in one, double negative, not enough responses

 b Overlapping responses, missing responses

2 85%

Page 18

1 a i 24 **ii** 21 **iii** 45 **iv** 48 **v** 72

 vi 49 **b i** 8 **ii** 7 **iii** 7 **iv** 3

 v 9 **vi** 8 **c i** 5 r 5 **ii** 9 r 1

 iii 7 r 1 **iv** 9 r 1 **v** 6 r 6

 vi 6 r 2

2 a i 14 **ii** 6 **iii** 15 **iv** 13 **v** 16

 vi 16

 b i 60 **ii** 30 **iii** 11 **iv** 3 **v** 6 **vi** 1

 c i $5 \times (6 + 1) = 35$

 ii $18 \div (2 + 1) = 6$

 iii $(25 - 10) \div 5 = 3$

 iv $(20 + 12) \div 4 = 8$

Page 19

1 a 5 hundred

 b Twenty-seven thousand, seven hundred and eight **c** 2 406 502

2 a i 60 **ii** 140 **iii** 50

 b i 700 **ii** 700 **iii** 1300

3 a 5830 **b** 2578

Page 20

1 a i 273 **ii** 324 **b i** 26 **ii** 34

 c 84 **d** 30

2 a 543 **b** 107 **c** 496 **d** 52

Page 21

1 a $\frac{7}{15}$ **b** $\frac{6}{9}$ **c** $\frac{3}{8}$ **d** $\frac{4}{6}$ **e** $\frac{4}{9}$

2 a i $\frac{5}{11}$ **ii** $\frac{3}{5}$ **iii** $\frac{6}{8}(\frac{3}{4})$

 b i $\frac{6}{9}(\frac{2}{3})$ **ii** $\frac{3}{5}$ **iii** $\frac{2}{13}$

3 a $\frac{3}{6}, \frac{5}{10}$ and $\frac{20}{40}$ **b** $\frac{4}{10}, \frac{8}{20}$ and $\frac{40}{100}$

 c b and d

Page 22

1 a i 18 **ii** 24 **iii** 20

 b i $\frac{2}{5}$ **ii** $\frac{1}{5}$ **iii** $\frac{3}{10}$ **iv** $\frac{3}{5}$ **v** $\frac{5}{7}$

2 a i $2\frac{2}{5}$ **ii** $3\frac{1}{4}$ **iii** $2\frac{2}{7}$ **iv** $2\frac{5}{8}$ **v** $2\frac{2}{3}$

 b i $\frac{8}{3}$ **ii** $\frac{23}{5}$ **iii** $\frac{20}{7}$ **iv** $\frac{9}{4}$ **v** $\frac{29}{8}$

Page 23

1 a i $\frac{4}{5}$ **ii** $\frac{2}{7}$ **iii** $\frac{5}{9}$ **iv** $\frac{1}{3}$

 b i $\frac{1}{3}$ **ii** $\frac{1}{5}$ **iii** $\frac{2}{5}$ **iv** $\frac{1}{3}$

 c i $1\frac{4}{9}$ **ii** $1\frac{2}{13}$ **iii** $1\frac{4}{11}$ **iv** $1\frac{1}{3}$

 d i $1\frac{1}{2}$ **ii** $1\frac{3}{5}$ **iii** $1\frac{1}{3}$ **iv** $1\frac{1}{2}$

2 a 8 **b** 20 **c** 25 **d** £225

 e 18 kg **f** 4 hours

 g $\frac{3}{4}$ of 20 = 15 ($\frac{2}{3}$ of 21 = 14)

 h $\frac{4}{7}$ of 63 = 36 ($\frac{7}{8}$ of 40 = 35)

Page 24

1 a $\frac{1}{5}$ **b** $\frac{3}{8}$ **c** $\frac{2}{9}$ **d** $\frac{3}{5}$ **e** $\frac{1}{36}$ **f** $\frac{1}{10}$

 g $\frac{1}{6}$ **h** $\frac{5}{12}$

2 a $\frac{2}{5}$ **b** $\frac{3}{4}$ **c** $\frac{1}{3}$

3 a $\frac{1}{3}$ **b** $\frac{1}{8}$ **c** £480

Page 25

1 a $0.3\dot{6}$ **b** $0.\dot{6}1\dot{5}$ **c** $0.3\dot{6}$ **d** $0.\dot{6}$

 e $0.1\dot{6}$ **f** $0.\dot{7}$

2 a $\frac{4}{5}$ **b** $\frac{13}{20}$ **c** $\frac{1}{8}$ **d** $2\frac{9}{20}$ **e** $\frac{1}{40}$ **f** $\frac{111}{125}$

3 a i 0.25 **ii** 0.05 **iii** $0.\dot{1}$

 b i $1\frac{1}{8}$ **ii** $1\frac{4}{5}$ **iii** $4\frac{1}{3}$

Page 26

1 a −£7 **b** −40 m **c** +8 **d** 24°F

 e +50 km **f** 2 km

2 a i < **ii** > **iii** < **b** −1, −4

 c i 5 **ii** −3 **iii** −3.5

Page 27

1 a −4 **b** −12 **c** 2 **d** −3 **e** 0

 f −4 **g** −6 **h** 5 **i** −14 **j** −4

 k 12 **l** −8 **m** 7 **n** −22 **o** −8

2 a −10 **b** −24 **c** 21 **d** 6

3 a −7 + 14 = +7 **b** 9 − 17 = −8

 c −6 + 14 − 5 = 3

Page 28

1 a i 6, 12, 18, 24, 30

 ii 13, 26, 39, 52, 65

 iii 25, 50, 75, 100, 125

 b i 250, 62, 78, 90, 108, 144, 96, 120

 ii 78, 90, 108, 144, 81, 96, 120, 333

 iii 250, 90, 85, 35, 120, 125

 iv 90, 108, 144, 81, 333

 v 250, 90, 120

 c Yes because 8 + 1 + 9 = 18 = 2 × 9

2 a i {1, 2, 3, 4, 6, 8, 12, 24}

 ii {1, 3, 5, 15}

 iii {1, 2, 5, 10, 25, 50}

 iv {1, 2, 4, 5, 8, 10, 20, 40}

 b {1, 2, 3, 4, 6, 8, 9, 12, 16, 18, 24, 36, 48, 72, 144}

Page 29

1 a i {1, 2, 3, 6, 9, 18}

 ii {1, 19}

 iii {1, 2, 4, 5, 10, 20}

 b 19 **c** 61, 79, 83, 17, 41, 29 **d** 2

2 a 36, 49, 64, 81, 100

 b i 121 **ii** 144 **iii** 169 **iv** 196

 v 225

 c   1 + 3 + 5 + 7 + 9 = 25

     1 + 3 + 5 + 7 + 9 + 11 = 36

 1 + 3 + 5 + 7 + 9 + 11 + 13 = 49

Page 30
1 a i 9 ii 8 iii 5
  b i +2, −2 ii +4, −4 iii +10, −10
  c i 24 ii 2.5 iii 6.1
2 a i 27 ii 64 iii 1000
  b i 4^5 ii 6^6 iii 10^4 iv 2^7
  c i 1024 ii 46 656 iii 10 000
    iv 128 d 64, 128, 256, 512, 1024

Page 31
1 a 800 b 64 c 250 d 300
  e 760 f 3250 g 6.4 h 0.028
  i 3.9 j 0.075 k 0.034 l 0.94
2 a 600 000 b 12 000 c 140 000
  d 200 e 20 f 50

Page 32
1 a 30 b 84 c 130 d 36 e 40 f 300
2 a $2^2 \times 5$ b $3^2 \times 5$ c 2^6
  d $2^3 \times 3 \times 5$
3 a 2^4 b $2 \times 3 \times 7$ c $2 \times 5 \times 7$
  d $2^3 \times 5^2$

Page 33
1 a i 30 ii 21 iii 39
  b The LCM is the product of the two
    numbers.
  c i 18 ii 40 iii 75
2 a 6 b 2 c 5 d 16 e 12 f 1
3 a 60 b 150 c 120

Page 34
1 a i 2^7 ii 2^9 iii 2^6
  b i 3^3 ii 3^4 iii 3^3
  c i x^9 ii x^9 iii x^{11}
  d i x^4 ii x^6 iii x^3
  e x^{n+m} f x^{n-m}
2 a 1 b 7 c 1 d 6
3 a $10x^7$ b $4x^4$

Page 35
1 a 552 b 3172 c 1953 d 2652
  e 6321 f 6890

Page 36
1 a 26 b 16 c 38 d 44 e 28
  f 49

Page 37
1 a 35 b £11.34
2 a i 2 ii 3 iii 1
  b i 2.3 ii 6.1 iii 15.9
  c i 3.45 ii 16.09 iii 7.63
  d i 4.974 ii 6.216 iii 0.008

Page 38
1 a i 57.8 ii 31.1 iii 36.5 iv 3.6
    v 5.83 vi 13.7
  b i 14 ii 14.1 iii 10.8 iv 2.89
    v 2.51 vi 4.65
  c i 106.03 ii 185.32 iii 58.48
    iv 1.32 v 0.091 vi 39.76

Page 39
1 a i $\frac{19}{28}$ ii $1\frac{5}{18}$ iii $6\frac{1}{15}$
  b i $\frac{13}{30}$ ii $\frac{2}{9}$ iii $\frac{7}{12}$
  c i $\frac{1}{6}$ ii $\frac{5}{14}$ iii $3\frac{17}{20}$
  d i $\frac{7}{10}$ ii $1\frac{3}{4}$ iii $1\frac{3}{5}$

Page 40
1 a i −15 ii +24 iii −35
  b i −4 ii −2 iii +3
2 a i 2 ii 3 iii 4
  b i 4 ii 0.8 iii 60
3 a 700 b 5 c 10 d 30−35

Page 41
1 a i 1:3 ii 5:6 iii 3:5
  b i 1:6 ii 8:1 iii 2:5
2 a £100 and £400 b 50 g and 250 g
  c £150 and £250 d 80 kg and 160 kg
3 a 56 b 30

Page 42
1 a 37.5 mph b 52.5 km
2 a 31.25 kg b 160
3 a Travel-size, 1.44 g/p compared to
    1.38 g/p
  b 95 out of 120 is 79.2% compared to
    62 out of 80, which is 77.5%.

Page 43
1 a i 0.3 ii 0.88 b i $\frac{9}{10}$ ii $\frac{8}{25}$
  c i 85% ii 15% d i $\frac{4}{5}$ ii $\frac{2}{25}$
  e i 0.625 ii 0.35 f i 36% ii 5%
2 a i 0.8 ii 0.07 iii 0.22
  b i 1.05 ii 1.12 iii 1.032
  c i 0.92 ii 0.85 iii 0.96

Page 44
1 a i £10.50 ii 19.2 kg
  b i £7.20 ii £52.80
2 a £168 b 66.24 kg
3 a i 740 ii 5% b 20%

Page 47
1 22 cm, 34 cm, 22 cm
2 15–18 cm², 12–15 cm²
3 i 15 m² ii 16 cm²

Page 48
1 a 35 cm² b 12.5 m²
2 a 46 c m² b 26 cm²

Page 49
1 a 20 cm² b 35 m²
2 a $37\frac{1}{2}$ cm² b 34 cm²

Page 50
1 Lengths are a, e and h
  Areas are c, g and i
  Volumes are b, d and f.

Page 51
1 2, 1, 4, 2, 1, 1
2 4, 6, 8, 2, 4, 2
3 9, 4, 4

Page 52
1 a 80° b 240°
2 a 38° b 116°

Page 53
1 $a = 75°$, $b = 72°$, $c = 36°$, $d = 60°$, $e = 22°$
2 Because it forms two triangles and the
  angles in each triangle add up to 180°,
  $2 \times 180° = 360°$.
3 $f = 90°$, $g = 90°$, $h = 117°$, $i = 63°$,
  $j = 109°$, $k = 71°$

Page 54
1 a 135° b 140°
2 a 45° b 40°
3 They always add up to 180°.

Page 55
1 a d b e c b d e e c f h

Page 56
1 a square, rhombus, kite
  b square, rectangle, parallelogram,
    rhombus
  c square, rhombus
  d square, rectangle, parallelogram,
    rhombus
  e square, rectangle, parallelogram,
    rhombus
  f trapezium g kite
2 a Yes, a square has all the properties of a
    rectangle.
  b Yes, a rhombus has all the properties
    of a parallelogram.
  c No, a kite does not have all sides
    equal.

Page 57
1 220°
2 a 270° (west) b 225° (south-west)
3 north-west
4 135° (south-east)

Page 58
1 a 44.0 cm b 37.7 cm c 8π cm
2 a 50.3 cm² b 706.9 cm² c 9π cm²

Page 59
1 a −2 °C b 26 mph c 25 g
2 a i 6–7 m ii 8–10 m
  b 1.2–1.5 kg

Page 60
1 a 150 cm b 1 m by 2 m c £2.50
2 Square-based pyramid and cube.

Page 61
1 a 2 cm by 3 cm by 4 cm b 24 cm³

2 a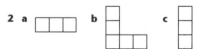

Page 62
1 a Yes b No c Yes
2 Any different tessellation using at least 6
  rectangles.

Page 63
1 a $\binom{5}{0}$ b $\binom{3}{4}$ c $\binom{5}{-4}$ d $\binom{0}{-4}$
2 a y-axis b $x = 3$

Page 64

1 **a** 90, anticlockwise about $\binom{0}{0}$
  b half-turn about $\binom{-1}{0}$
2 **a i** 2 **b i** $\frac{2}{3}$
  b ii / **a ii**

Page 65

1 2 55°
3 40° 65° 4 60°

Page 66

1–3 Self checking.

Page 67

1 Self checking.
2
•Edinburgh
•London

Page 68

1 **a** 1.2 m **b** 3.5 kg **c** 2.3 l
  d 4.5 cm
2 **a** 55 lbs **b** 36 l **c** 90 cm
  d 75 miles

Page 69

1 **a** 2 m³ **b** 3 m² **c** 5 000 000 cm³
  d 400 mm²
2 **a** 52 cm² **b** 24 cm³

Page 70

1 **a** 6.25 g/cm³ **b** 27 kg **c** 0.3 m³
2 **a i** 24 cm² **ii** 240 cm³ **b** 42 cm³
3 160π cm³

Page 71

1 3.9 m

Page 74

1 **a** $r - p$ **b** $7 + x$ **c** ab
  d $\frac{t}{2}$ **e** $n + m$
2 **a** $8x$ **b** $9x + 5$ **c** $9w - 6k$
  d $2x^2 - 2z$
3 **a** $12t$ **b** $12n^3$ **c** $30m^2n$ **d** $-12x^3y^4$

Page 75

1 **a** $3x + 15$ **b** $5y - 10$ **c** $6x + 3y$
  d $n^2 - 7n$ **e** $10m^2 + 15m$
  f $6p^3 - 9p^2q$
2 **a** $6x + 6$ **b** $3m + 22$ **c** $4n^2 - 5n$
  d $3x + 27$ **e** $4x^2 - 7xy$
  f $10x + 38y$
3 **a** $5(n + 2m)$ **b** $3x(2x - 3)$
  c $m(5n + 6)$ **d** $4xy(x + 3y)$
  e $2x(y + 3x)$
  f $2ab(a - 4 + 3b)$

Page 76

1 **a** $x^2 + 2x - 3$ **b** $m^2 - 4m - 12$
  c $n^2 - n - 2$ **d** $x^2 + 6x + 5$
  e $x^2 - 6x + 9$ **f** $x^2 + 10x + 21$
2 **a** $x^2 + 6x + 9$ **b** $x^2 - 4x + 4$
  c $x^2 - 1$ **d** $m^2 - 4$
3 **a** –6 **b** 29 **c** 24

Page 77

1 **a** 3 **b** –1 **c** 12 **d** $\frac{1}{2}$ **e** 13 **f** 2
  g 13 **h** 28 **i** $\frac{1}{2}$

Page 78

1 **a** 9 **b** 6 **c** –2 **d** 4 **e** $2\frac{1}{2}$
  f $-1\frac{1}{2}$
2 **a** 7 **b** 2 **c** 3 **d** –1 **e** $\frac{1}{5}$ **f** $-2\frac{1}{2}$
3 **a** 9 **b** –3 **c** 5 **d** $6\frac{1}{2}$

Page 79

1 **a** 4 **b** 6
2 **a** $5x + 7 = 22$ **b** 3
3 **a** $\frac{x}{2} + 7 = x + 6$ **b** 2

Page 80

1 4.8
2 2.6

Page 81

1 Formulae: **a** and **c**; identities: **d** and **h**;
  expressions **f** and **g**, equations **b** and **e**
2 **a** $x = \frac{T}{4}$ **b** $x = \frac{y - 3}{2}$ **c** $x = P - t$
  d $x = 5y$ **e** $x = \frac{A}{m}$ **f** $x = \frac{S}{2\pi}$

Page 82

1 **a** $x < 3$ **b** $x > 1$ **c** $x \geqslant 18$
  d $x \leqslant -1$ **e** $x > -10$ **f** $x \leqslant -1$
2 **a** $x > 1$ **b** $x \leqslant 3$
3 **a** $x < 2$ **b** $x \geqslant -1$ **c** –1, 0, 1

Page 83

1 **a** £45 **b** 150 units
2 **a** 20 minutes
  b The car was stationary.
  c B to C **d** 45 km/h

Page 84

1 B(3, 0), C(0, 1), D(–2, 4), F(–2, 0),
  G(–4, –1), I(1, –3)
2 **a** –5, –3, –1
  b
3

Page 85

1 **a** A and C **b** A and B
2 **a**
  b

Page 86

1 **a** 1, –1, –1
  b

Page 87

1 **a** +1.4, –1.4 **b** 4.4 and –1.4

Page 88

1 **a** 12345 × 9 + 6 = 111111,
    123456 × 9 + 7 = 1111111
  b 123456 × 8 + 6 = 987654,
    1234567 × 8 + 7 = 9876543
  c 5 × 3 × 37 = 555, 6 × 3 × 37 = 666
  d 66667 × 66667 = 4444488889,
    666667 × 666667 = 444444888889
2 **a i** 39, 45 **ii** goes up in 6s
  b i 96, 192 **ii** doubles
  c i 78, 74 **ii** down in 4s
  d i 59, 75 **ii** up 2, 4, 6, 8 …
  e i –2, –5 **ii** down in 3s

Page 89

1 **a i** 3, 7, 11 **ii** 399
  b i 3, 6, 10 **ii** 20 100
3 **a** $5n + 1$ **b** $8n - 5$ **c** $3n + 6$

Page 90

1 **a** 200 **b** 199 **c** 10 000
  d 28, 36, 45, 55 **e** 128, 256, 512, 1024
  f 23, 29, 31, 37, 41
2 **a** 41 **b** $4n + 1$

Index

Collins Revision

GCSE Foundation Maths

Exam Practice Workbook

FOR EDEXCEL A

FOR AQA A

FOR AQA B

Keith Gordon

Statistical representation

1 Zeke did a survey of the number of passengers in some cars passing the school.

The results are shown in the table.

Number of passengers	Tally	Frequency
1	ⅢⅠ ⅢⅠ ⅢⅠ ⅢⅠ Ⅰ	21
2	ⅢⅠ ⅢⅠ Ⅲ	
3		8
4	ⅢⅠ Ⅰ	
5	Ⅱ	

a Complete the frequency column. **[1 mark]**

b Complete the tally column. **[1 mark]**

c How many cars were surveyed altogether?

_____ **[1 mark]**

d Explain why the total number of passengers in all the cars was 105.

_____

_____ **[2 marks]**

2 The pictogram shows the number of letters delivered to a company during five days.

⊠ represents 20 letters.

Day		Number of letters
Monday	⊠ ⊠ ⊠ ▷	65
Tuesday	⊠ ⊠ ⊠ ⊠ ⊠	
Wednesday	⊠ ⊠ ⊠ ◸	
Thursday		40
Friday		35

a Complete the 'number of letters' column. **[1 mark]**

b Complete the pictogram column. **[1 mark]**

c How many letters were delivered altogether during the week?

_____ **[1 mark]**

This page tests you on • statistics • collecting data • pictograms

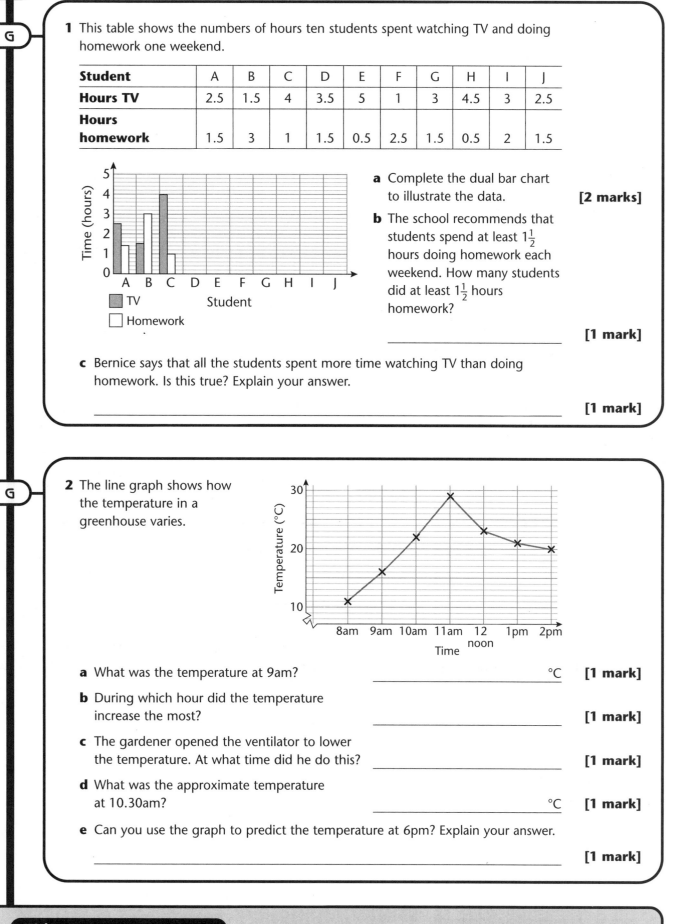

1 This table shows the numbers of hours ten students spent watching TV and doing homework one weekend.

Student	A	B	C	D	E	F	G	H	I	J
Hours TV	2.5	1.5	4	3.5	5	1	3	4.5	3	2.5
Hours homework	1.5	3	1	1.5	0.5	2.5	1.5	0.5	2	1.5

a Complete the dual bar chart to illustrate the data. **[2 marks]**

b The school recommends that students spend at least $1\frac{1}{2}$ hours doing homework each weekend. How many students did at least $1\frac{1}{2}$ hours homework?

_____ **[1 mark]**

c Bernice says that all the students spent more time watching TV than doing homework. Is this true? Explain your answer.

_____ **[1 mark]**

2 The line graph shows how the temperature in a greenhouse varies.

a What was the temperature at 9am? _____ °C **[1 mark]**

b During which hour did the temperature increase the most? _____ **[1 mark]**

c The gardener opened the ventilator to lower the temperature. At what time did he do this? _____ **[1 mark]**

d What was the approximate temperature at 10.30am? _____ °C **[1 mark]**

e Can you use the graph to predict the temperature at 6pm? Explain your answer.

_____ **[1 mark]**

This page tests you on • bar charts • line graphs

Averages

1 The table shows the numbers of passengers in some cars.

Number of passengers	Frequency
1	23
2	15
3	5
4	4
5	3

a How many cars were in the survey?

_____ [1 mark]

b What is the modal number of passengers per car?

_____ [1 mark]

c What is the median number of passengers per car?

_____ [1 mark]

d Cars with two or more passengers can use a 'car pool' lane on the motorway. What percentage of these cars could use the car pool lane?

_____ [1 mark]

F

2 a For the data 51, 74, 53, 74, 76, 58, 68, 51, 70 and 65 work out:

i the mean

_____ [1 mark]

ii the range

_____ [1 mark]

b A football team of 11 players has a mean weight of 84 kg.

i How much do the 11 players weigh altogether?

_____ [1 mark]

ii When the three substitutes are included, the 14 players have a mean weight of 87 kg. What is the mean weight of the three substitutes?

_____ [2 marks]

E

This page tests you on   • mode • median • mean

F

1 The bar chart shows the numbers of spoonfuls of sugar that a group of workmen take in their morning tea.

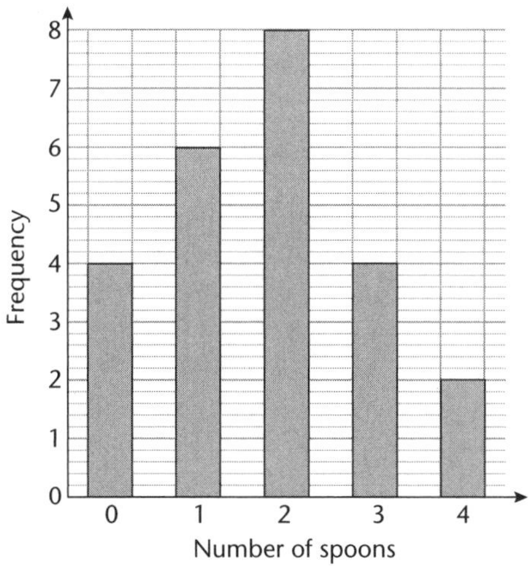

a How many workmen are there?

_____ **[1 mark]**

b What is the modal number of spoons?

_____ **[1 mark]**

c What is the median number of spoons?

_____ **[1 mark]**

d What is the mean number of spoons?

_____ **[2 marks]**

e The workmen buy a kilogram of sugar. This is enough for 400 spoons of sugar. Will they have enough sugar for five days if they each have two cups of tea a day? Explain your answer fully.

_____ **[1 mark]**

D

2 Ten workmates went ten-pin bowling.

a Their scores for the first game were: 87 123 121 103 93 231 145 46 65 46
Work out:

i the modal score _____ **[1 mark]**

ii the median score _____ **[1 mark]**

iii the mean score. _____ **[1 mark]**

b i Explain why the mode would not be a good average to use.

_____ **[1 mark]**

ii Explain why the mean would not be a good average to use.

_____ **[1 mark]**

c In the second game, the modal score was 105. Does this mean the players increased their overall scores? Explain your answer.

_____ **[1 mark]**

d In the third game, the mean was 120.

i Does this mean the players increased their overall scores? Explain your answer.

_____ **[1 mark]**

ii Can you tell whether the median for the third game was higher or lower than for the first game? Explain your answer.

_____ **[1 mark]**

This page tests you on • range • which average to use

Arranging data

1 The table shows the number of cars per house on a housing estate of 100 houses.

Number of cars	Number of houses
0	8
1	23
2	52
3	15
4	2

Work out:

a the modal number of cars

_____ [1 mark]

b the median number of cars

_____ [1 mark]

c the mean number of cars.

_____ [2 marks]

D

2 The table shows the scores of 200 boys in a mathematics exam.

The frequency polygon shows the scores of 200 girls in the same examination.

Mark, x	Frequency, f
$40 < x \leqslant 50$	27
$50 < x \leqslant 60$	39
$60 < x \leqslant 70$	78
$70 < x \leqslant 80$	31
$80 < x \leqslant 90$	13
$90 < x \leqslant 100$	12

a Work out the mean mark for the boys' scores.

_____ [2 marks]

b On the same graph as the girls frequency polygon, draw the frequency polygon for the boys' scores. [1 mark]

c Who did better in the test, the boys or the girls? Give reasons for your answer.

_____

_____ [1 mark]

C

This page tests you on • frequency tables • grouped data

D

1 These are the weights of 20 guinea pigs, in grams, rounded to the nearest 10 grams.

130   90   220   210   190   130   160   110   230   90

80   120   130   240   180   150   70   220   240   130

a Using the key 1 | 3 to represent 130 grams, put the data into the stem-and-leaf diagram.

0 |
1 |
2 |

[2 marks]

Key 1 | 3 represents 130 grams

b Using the stem-and-leaf diagram, or otherwise, write down:

i the modal weight

_____ grams   **[1 mark]**

ii the median weight

_____ grams   **[1 mark]**

iii the range of the weights.

_____ grams   **[1 mark]**

C

2 A teacher recorded how many times her students were late during a term.

The stem-and-leaf diagram shows the data.

12 students were **never late**.

0	2	3	4	4	5	6	7
1	3	5	8	9	9		
2	0	1	4	5			
3	2						
5	1						

Key 1 | 7 represents 17 times late

a How many students there are in the form altogether?

_____   **[1 mark]**

b Work out the mean number of times late for the whole form.

_____   **[2 marks]**

This page tests you on • stem-and-leaf diagrams

Probability

1 a State whether each of the following events is *impossible, very unlikely, unlikely, evens, likely, very likely* or *certain*.

G

   i you walking on the moon tomorrow

   _____ **[1 mark]**

   ii getting a six when a regular dice is thrown

   _____ **[1 mark]**

   iii tossing a coin and scoring a head

   _____ **[1 mark]**

   iv someone in the class going abroad for their holidays

   _____ **[1 mark]**

b On the probability scale, put a numbered arrow to show approximately the probability of each of the following outcomes of events happening.

```
|--------------------------|--------------------------|
0                          ½                          1
```

   i The next car you see driving down the road will only have the driver inside. **[1 mark]**
   ii Someone in the class had porridge for breakfast. **[1 mark]**
   iii Picking a red card from a well shuffled pack of cards. **[1 mark]**
   iv Throwing a number less than seven with a regular dice. **[1 mark]**

2 A bag contains 20 coloured balls. Twelve are red, five are blue and the rest are white. A ball is taken from the bag at random.

E

   a What is the probability that the ball is:
   i red

   _____ **[1 mark]**

   ii pink

   _____ **[1 mark]**

   iii blue or white?

   _____ **[1 mark]**

   b Some more white balls are added to the bag so that the probability of getting a red ball is now $\frac{1}{2}$. How many red balls were added?

   _____ **[1 mark]**

This page tests you on • the probability scale • calculating probabilities

1 The probability that a milkman delivers the wrong sort of milk to a house is $\frac{3}{50}$.

a What is the probability that he delivers the correct sort of milk to a house?

_____ **[1 mark]**

b He delivers milk to 500 houses a day. Estimate the number of houses that get the wrong milk.

_____ **[1 mark]**

2 There are 900 squares on a 'Treasure map' at the school Summer Fayre.

One of the squares contains the treasure.

The Rogers decide to buy some squares.

Mr Rogers buys five squares, Mrs Rogers buys ten squares and their two children, Amy and Ben, buy two squares each.

a Which member of the family has the best chance of winning?

Explain your answer.

_____ **[1 mark]**

b What is the probability that Mr Rogers wins the treasure?

_____ **[1 mark]**

c What is the probability that one of the children wins the treasure?

_____ **[1 mark]**

d If the family put all their squares together, what is the probability that they will win the treasure?

_____ **[1 mark]**

3 John makes a dice and weights one side with a piece of sticky gum. He throws it 120 times. The table shows the results.

Score	1	2	3	4	5	6
Frequency	18	7	22	21	35	17
Relative frequency						

a Fill in the relative frequency row. Give your answers to 2 decimal places. **[2 marks]**

b Which side did John stick the gum on? Explain how you can tell.

_____ **[1 mark]**

This page tests you on
- probability of 'not' an event
- addition rule for mutually exclusive outcomes
- relative frequency

1 Pete's Café has a breakfast deal.

Three-item breakfast! Only £1		
Choose one of:	**Choose one of:**	**Choose one of:**
sausage or bacon	egg or hash browns	beans or toast

    a There are eight possible 'three-item breakfast' combinations, for example,
       sausage, **egg**, **beans** or **sausage**, **egg**, **toast**.

       List all the other possible combinations.

       _____   _____

       _____   _____

       _____   _____

       _____   _____ **[2 marks]**

    b Fred tells his friend, 'I'll have what you are having.'

       What is the probability that Fred gets bacon and eggs with his breakfast?

       _____ **[1 mark]**

E

2 The sample space diagram shows the
outcomes when a coin and a regular dice
are thrown at the same time.

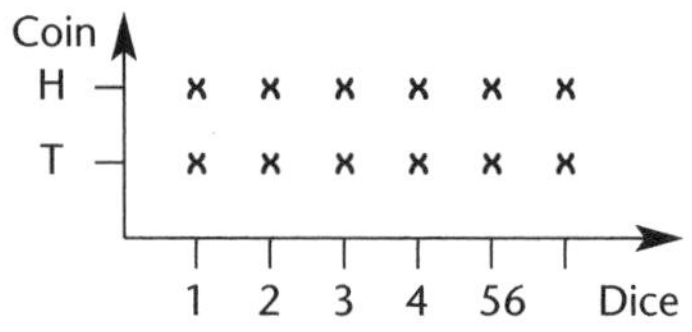

    a How many possible outcomes are there
       when a coin and a dice are thrown together?

       _____ **[1 mark]**

    b When a coin and a dice are thrown together what is the probability that:

       i the coin shows tails and the dice shows an even number

       _____ **[1 mark]**

       ii the coin shows heads and the dice shows a square number?

       _____ **[1 mark]**

    c Alicia and Zeek play a game with the coin
       and dice. If the coin lands on a head the
       score on the dice is doubled. If the coin
       lands on a tail 1 is subtracted from the
       score on the dice.

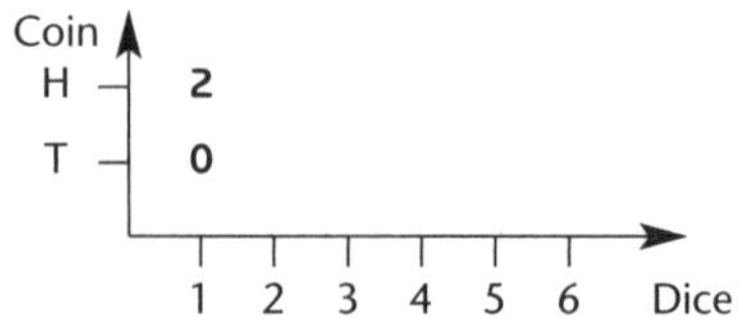

       i Complete the sample space diagram to show all possible scores. **[2 marks]**

       ii What is the probability of an odd score?

       _____ **[1 mark]**

D

This page tests you on   • **combined events**

C

1 A bag contains 30 balls that are either red or white. The ratio of red balls to white balls is 2 : 3.

    a Zoe says that the probability of picking a red ball at random from the bag is $\frac{2}{3}$.

       Explain why Zoe is wrong.

       _____

       _____ **[1 mark]**

    b How many red balls are there in the bag?

       _____ **[1 mark]**

    c A ball is taken from the bag at random, its colour noted and then it is replaced.

       This is done 200 times. How many of the balls would you expect to be red?

       _____ **[1 mark]**

E

2 The two-way table shows the numbers of male and female teachers in four school departments.

	Male	Female
Mathematics	7	5
Science	11	7
RE	1	3
PE	3	3

    a How many male teachers are there altogether?

       _____ **[1 mark]**

    b Which subject has equal numbers of male and female teachers?

       _____ **[1 mark]**

    c Nuna says that science is a more popular subject than mathematics for female teachers.

       Explain why Nuna is wrong.

       _____

       _____ **[1 mark]**

    d What is the probability that a teacher chosen at random from the table will be a female teacher of mathematics or science?

       _____

       _____ **[1 mark]**

This page tests you on   • expectation   • two-way tables

Pie charts

1 The table shows the results of a survey of 60 students, to find out what they do for lunch.

Lunch arrangement	Frequency	Angle
Use canteen	22	
Have sandwiches	18	
Go home	12	
Go to shopping centre	8	

Tom is going to draw a pie chart to show the data.

a Complete the column for the angle for each sector.   **[2 marks]**

b Draw a fully labelled pie chart below.

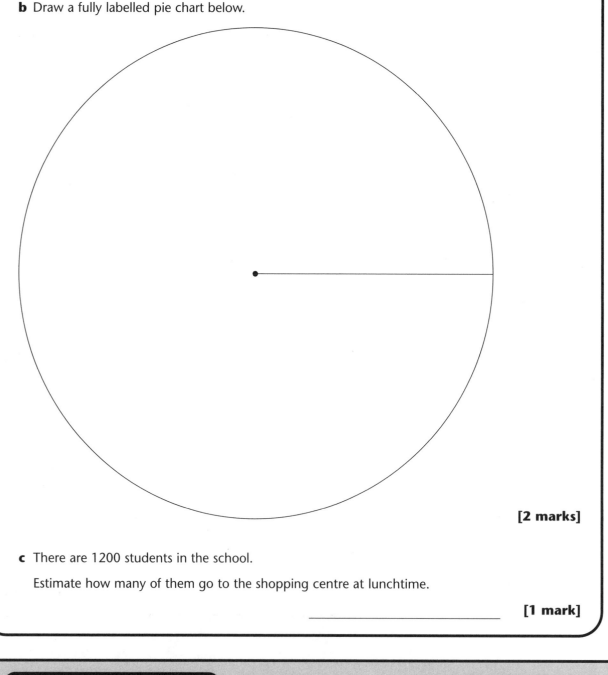

[2 marks]

c There are 1200 students in the school.

Estimate how many of them go to the shopping centre at lunchtime.

_____   **[1 mark]**

This page tests you on   • pie charts

Scatter diagrams

1 A delivery driver records the distances and times for deliveries.

The table shows the results.

Delivery	Distance (km)	Time (minutes)
A	12	30
B	16	42
C	20	55
D	8	15
E	18	40
F	25	60
G	9	45
H	15	32
I	20	20
J	14	35

a Plot the values on the scatter diagram.

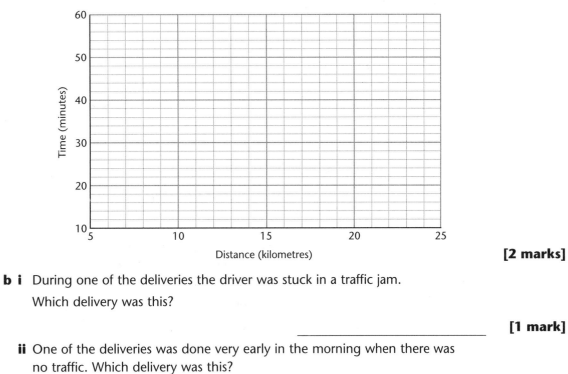

[2 marks]

b i During one of the deliveries the driver was stuck in a traffic jam.
Which delivery was this?

_____ [1 mark]

ii One of the deliveries was done very early in the morning when there was
no traffic. Which delivery was this?

_____ [1 mark]

c Ignoring the two values in **b**, draw a line of best fit through the rest of
the data. [1 mark]

d Under normal conditions, how long would you expect a delivery of 22 kilometres
to take?

_____ [1 mark]

This page tests you on • scatter diagrams • correlation • line of best fit

Surveys

1 Danny wanted to investigate the hypothesis:

> ***Girls spend more time on mathematics coursework than boys do***.

a Design a two-way table that will help Danny collect data.

[2 marks]

b Danny collected data from 30 boys and 10 girls.

He found that, on average, the boys spent 10 hours and the girls spent 11 hours on their mathematics coursework.

Does this prove the hypothesis? Give reasons for your answer.

_____ **[1 mark]**

D

2 This graph shows the price index of petrol from 2000 to 2007. In 2000 the price of a litre of petrol was 60p.

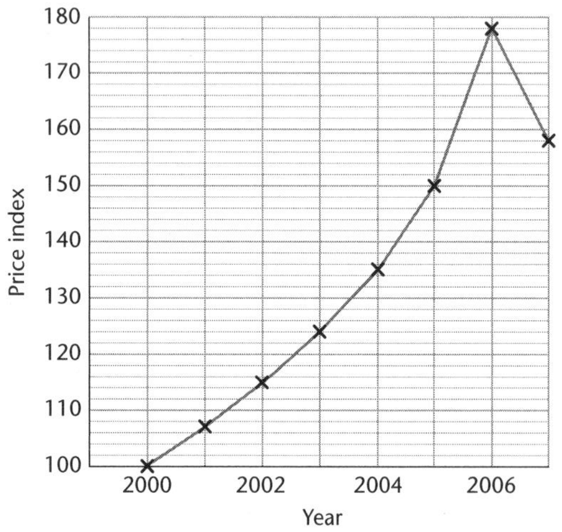

a Use the graph to estimate the price of a litre of petrol in 2007.

_____ **[1 mark]**

b A war in Iraq caused an oil crisis. Which year do you think this happened?

Explain how you can tell.

_____

_____

_____ **[1 mark]**

c Over the same period, the index for the cost of living rose by 38%.

Is the cost of petrol increasing faster or more slowly than the cost of living?

Explain your answer.

_____ **[1 mark]**

C

This page tests you on • surveys • social statistics

Handling data checklist

I can...

- ☐ draw and read information from bar charts, dual bar charts and pictograms
- ☐ find the mode and median of a list of data
- ☐ understand basic terms such as 'certain', 'impossible', 'likely'

You are working at (Grade G) level.

- ☐ work out the total frequency from a frequency table and compare data in bar charts
- ☐ find the range of a set of data
- ☐ find the mean of a set of data
- ☐ understand that the probability scale runs from 0 to 1
- ☐ calculate the probability of events with equally likely outcomes
- ☐ interpret a simple pie chart

You are working at (Grade F) level.

- ☐ read information from a stem-and-leaf diagram
- ☐ find the mode, median and range from a stem-and-leaf diagram
- ☐ list all the outcomes of two independent events and calculate probabilities from lists or tables
- ☐ calculate the probability of an event not happening when the probability of it happening is known
- ☐ draw a pie chart

You are working at (Grade E) level.

- ☐ draw an ordered stem-and-leaf diagram
- ☐ find the mean of a frequency table of discrete data
- ☐ find the mean from a stem-and-leaf diagram
- ☐ predict the expected number of outcomes of an event
- ☐ draw a line of best fit on a scatter diagram
- ☐ recognise the different types of correlation
- ☐ design a data collection sheet
- ☐ draw a frequency polygon for discrete data

You are working at (Grade D) level.

- ☐ find an estimate of the mean from a grouped table of continuous data
- ☐ draw a frequency diagram for continuous data
- ☐ calculate the relative frequency of an event from experimental data
- ☐ interpret a scatter diagram
- ☐ use a line of best fit to predict values
- ☐ design and criticise questions for questionnaires.

You are working at (Grade C) level.

Basic number

1 a Write down the answer to each calculation.

G

i 8×9

_____ [1 mark]

ii 40×7

_____ [1 mark]

iii $240 \div 3$

_____ [1 mark]

b George has four cards with a number written on each of them.

4 **8** **2** **7**

i He uses the cards to make a multiplication statement.

☐ $\times$ ☐ $=$ ☐ ☐

What is the multiplication statement?

_____ [1 mark]

ii He uses the cards to make a division statement.

☐ ☐ $\div$ ☐ $=$ ☐

What is the division statement?

_____ [1 mark]

2 Two students work out the following calculation:

D

$2 + 4^2 \div 8$

Sammi gets an answer of 2.25. Ross gets an answer of 4.5.

Both of these answers are wrong.

a What is the answer to $2 + 4^2 \div 8$?

_____ [1 mark]

b Put brackets in the following calculations to make them true.

i $2 + 4^2 \div 8 = 2.25$ [1 mark]

ii $2 + 4^2 \div 8 = 4.5$ [1 mark]

This page tests you on   • times tables   • order of operations and BODMAS

G

1 Work these out.

a 2 5 7 6
  + 1 0 8 3
  _____

b   1 2 9
  + 6 7 3 5
  _____

c 78 + 2054 − 362

  _____ **[1 mark each]**

G

2 a When England won the world cup at Wembley in 1964 the attendance was given as 96 924.

  i Write 96 924 in words.

  _____ **[1 mark]**

  ii Round 96 924 to the nearest 100. _____ **[1 mark]**

  iii Round 96 924 to the nearest 1000. _____ **[1 mark]**

  b When Italy won the world cup in Berlin in 2006, the attendance was given as 69 000 rounded to the nearest hundred.

  i What is the value of the digit 9 in 69 000? _____ **[1 mark]**

  ii What is the smallest value the attendance could have been? _____ **[1 mark]**

  iii What is the largest value the attendance could have been? _____ **[1 mark]**

  iv About one third of the spectators were Italian.

  Approximately how many spectators were Italian? _____ **[1 mark]**

F

3 Work these out.

a 3 0 7 6
  − 2 1 7 8
  _____

b 6 9 0 3
  − 3 7 2 5
  _____

c 86 + 1623 − 484

  _____ **[1 mark each]**

F

4 Farmer Bill has 1728 sheep on his farm. Farmer Jill has 589 sheep on her farm.

  a How many more sheep does farmer Bill have than farmer Jill?

  _____ **[1 mark]**

  b Farmer Jill sells all of her sheep to farmer Bill. How many does he have now?

  _____ **[1 mark]**

This page tests you on • place value • rounding
• column addition and subtraction

1 Work these out.

    a   7 6
            b   **c** 54 × 7   **d** 384 ÷ 8

        ×   4   6)‾1‾5‾6‾

    ‾‾‾‾‾‾   ‾‾‾‾‾‾‾   ‾‾‾‾‾‾‾ **[1 mark each]**

G

2 a Mary buys four cans of cola at 68p per can.

    i   How much do the four cans cost altogether?

                                      ‾‾‾‾‾‾‾‾‾‾‾‾‾‾‾‾‾‾‾‾‾ **[1 mark]**

    ii She pays with a £5 note. How much change does she get?

                                        ‾‾‾‾‾‾‾‾‾‾‾‾‾‾‾‾‾‾‾‾‾ **[1 mark]**

  b In the school hall, there are 24 chairs in each row.

    There are 30 rows of chairs.

    How many chairs are there in total?

                                      ‾‾‾‾‾‾‾‾‾‾‾‾‾‾‾‾‾‾‾‾‾ **[1 mark]**

  c Year 10 has 196 students in seven forms.

    Each form has the same number of students.

    How many students are there in each form?

                                      ‾‾‾‾‾‾‾‾‾‾‾‾‾‾‾‾‾‾‾‾‾ **[1 mark]**

G

3 Show the calculation you need to do to work out each answer.

Then calculate the answer.

  a How much change do I get from £20 if I spend £12.85?

    ‾‾‾

    ‾‾‾ **[1 mark]**

  b I buy three ties at a total cost of £14.55. What is the price of each tie?

    ‾‾‾

    ‾‾‾ **[1 mark]**

  c Cartons of eggs contain 12 eggs. How many eggs will there be in nine cartons?

    ‾‾‾

    ‾‾‾ **[1 mark]**

F

This page tests you on   • multiplying and dividing by single-digit numbers
                                           • problems in words

Fractions

1 a Which two of these fractions are equivalent to $\frac{3}{4}$?

$\frac{9}{12}$ $\frac{6}{7}$ $\frac{8}{20}$ $\frac{60}{80}$

_____ and _____ **[1 mark]**

b Shade $\frac{3}{4}$ of this shape.

[1 mark]

c i What fraction of this shape is shaded?

_____ **[1 mark]**

ii What fraction is not shaded?

_____ **[1 mark]**

2 a i Shade $\frac{2}{9}$ of this shape.

[1 mark]

ii Shade $\frac{1}{3}$ of this shape.

[1 mark]

b Use your answer to part (a) to write down the answer to:

$\frac{2}{9} + \frac{1}{3} =$

_____ **[1 mark]**

3 The ratio of grey squares to white squares in this shape is 2:3.

a Zoe says, 'That means the grey squares must be $\frac{2}{3}$ of the shape.'

Explain why Zoe is wrong.

_____ **[1 mark]**

b Write down **two** other fractions that are equivalent to $\frac{2}{5}$.

_____ and _____ **[1 mark]**

This page tests you on
- fractions of a shape
- adding and subtracting simple fractions
- equivalent fractions

1 a Shade squares in each of these diagrams so that $\frac{1}{4}$ of each diagram is shaded.

i

ii

iii

iv

[2 marks]

b Fill in the boxes to make the following fractions equivalent.

i $\frac{3}{7} = \frac{\square}{28}$

ii $\frac{5}{8} = \frac{20}{\square}$

iii $\frac{15}{18} = \frac{5}{\square}$

iv $\frac{\square}{3} = \frac{16}{24}$

[1 mark each]

c Cancel the following fractions, giving each answer in its simplest form.

i $\frac{16}{28} = $ _____

ii $\frac{9}{15} = $ _____

iii $\frac{12}{30} = $ _____

iv $\frac{21}{28} = $ _____

[1 mark each]

d Put the following fractions in order, with the smallest first.

$\frac{7}{10}$ $\frac{4}{5}$ $\frac{13}{20}$

_____ **[1 mark]**

G

2 a Change the following top-heavy fractions into mixed numbers.

i $\frac{9}{5} = $ _____

ii $\frac{17}{7} = $ _____

iii $\frac{21}{8} = $ _____

iv $\frac{31}{4} = $ _____

[1 mark each]

b Change the following mixed numbers into top-heavy fractions.

i $1\frac{6}{11} = $ _____

ii $1\frac{3}{8} = $ _____

iii $2\frac{1}{3} = $ _____

iv $4\frac{3}{5} = $ _____

[1 mark each]

F

This page tests you on
- equivalent fractions and cancelling
- top-heavy fractions and mixed numbers

E

1 a Fill in the boxes to make the following fractions equivalent.

i $\frac{3}{10} = \frac{\Box}{20}$ ii $\frac{5}{8} = \frac{\Box}{16}$ iii $\frac{1}{3} = \frac{\Box}{6}$ iv $\frac{3}{4} = \frac{\Box}{12}$ **[1 mark each]**

b Use the answers to part **a** to work these out.

Cancel the answer to its simplest form.

i $\frac{1}{20} + \frac{3}{10}$ _____ **[1 mark]**

ii $\frac{3}{16} + \frac{5}{8}$ _____ **[1 mark]**

iii $\frac{1}{6} + \frac{1}{3}$ _____ **[1 mark]**

iv $\frac{1}{12} + \frac{3}{4}$ _____ **[1 mark]**

c Use the answers to part **a** to work these out.

Cancel the answer to its simplest form.

i $\frac{17}{20} - \frac{3}{10}$ _____ **[1 mark]**

ii $\frac{15}{16} - \frac{5}{8}$ _____ **[1 mark]**

iii $\frac{5}{6} - \frac{1}{3}$ _____ **[1 mark]**

iv $\frac{11}{12} - \frac{3}{4}$ _____ **[1 mark]**

F

2 a On MacDonald's Farm there are a total of 245 hectares.

$\frac{2}{5}$ of the land is planted for crops, the rest is used for animals.

i What fraction of the land is used for animals?

_____ **[1 mark]**

ii How many hectares are used for crops?

_____ **[1 mark]**

b MacDonald has 120 sheep.

$\frac{3}{4}$ of the sheep each gives birth to two lambs, the rest give birth to one lamb.

How many lambs are born altogether?

_____ **[2 marks]**

c MacDonald has 220 cows. He sells $\frac{1}{4}$ of them.

How many cows has he got left after that?

_____ **[2 marks]**

This page tests you on
- adding and subtracting fractions
- finding a fraction of a quantity

1 a Work these out, giving each answer in its simplest form.

i $\frac{2}{3} \times \frac{6}{11}$ _____ [1 mark]

ii $\frac{3}{8} \times \frac{4}{9}$ _____ [1 mark]

iii $\frac{5}{6} \times \frac{3}{20}$ _____ [1 mark]

iv $\frac{9}{10} \times \frac{5}{6}$ _____ [1 mark]

E

b Work out each of these and give the answer as a mixed number in its simplest form.

i $4 \times \frac{3}{8}$ _____ [1 mark]

ii $5 \times \frac{3}{10}$ _____ [1 mark]

iii $6 \times \frac{1}{3}$ _____ [1 mark]

iv $8 \times \frac{3}{4}$ _____ [1 mark]

2 a In a school there are 1500 students. 250 of the students are in Year 7.

What fraction of the students are in Year 7?

_____ [2 marks]

E

b There are 120 girls in Year 7. 40 of the girls are left-handed.

What fraction of the girls are left-handed?

_____ [2 marks]

3 a Frank earns £18 000 per year. He pays $\frac{3}{20}$ of his pay in tax.

How much tax does he pay?

_____ [2 marks]

E

b Packets of washing powder normally contain 1.2 kg.

A special offer pack contains $\frac{1}{6}$ more than a normal pack.

How much does the special offer pack contain?

_____ [2 marks]

This page tests you on
• multiplying fractions
• one quantity as a fraction of another
• problems in words

Rational numbers and reciprocals

D

1 a Write each fraction as a decimal. Give the answer as a terminating decimal or a recurring decimal as appropriate.

i $\frac{7}{40}$ _____ [1 mark]

ii $\frac{11}{15}$ _____ [1 mark]

iii $\frac{5}{6}$ _____ [1 mark]

iv $\frac{9}{50}$ _____ [1 mark]

b $\frac{1}{9} = 0.1111...$ $\quad$ $\frac{2}{9} = 0.2222...$

Use this information to write down:

i $\frac{4}{9}$ _____ [1 mark]

ii $\frac{5}{9}$ _____ [1 mark]

c $\frac{1}{11} = 0.0909...$ $\quad$ $\frac{2}{11} = 0.1818...$

Use this information to write down:

i $\frac{3}{11}$ _____ [1 mark]

ii $\frac{6}{11}$ _____ [1 mark]

C

2 a Write down the reciprocal of each of these numbers.

Give your answer as a fraction or a mixed number.

i 10 _____ [1 mark]

ii $\frac{3}{4}$ _____ [1 mark]

b Write each of the answers to **a** as a terminating or a recurring decimal.

i _____ [1 mark]

ii _____ [1 mark]

c Work out the reciprocal of each number.

i 1.25 _____ [1 mark]

ii 2.5 _____ [1 mark]

iii 5 _____ [1 mark]

d The sequence 1.25, 2.5, 5, 10, ... is formed by doubling each term to find the next term.

Using your answer to part **c**, explain how you know the reciprocal of 40 is 0.025 without doing the calculation 1 ÷ 40.

_____

_____ [2 marks]

This page tests you on
- rational numbers
- converting terminating decimals into fractions
- finding reciprocals

Negative numbers

1 These maps show the maximum and minimum temperatures in five towns during a six-month period.

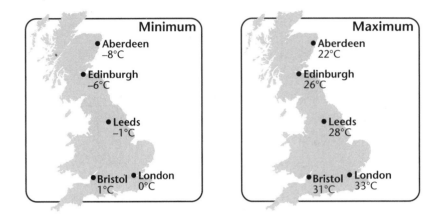

Minimum
- Aberdeen –8°C
- Edinburgh –6°C
- Leeds –1°C
- Bristol 1°C • London 0°C

Maximum
- Aberdeen 22°C
- Edinburgh 26°C
- Leeds 28°C
- Bristol 31°C • London 33°C

a Which town had the lowest minimum temperature?

_____ [1 mark]

b What is the difference between the lowest and highest **minimum** temperatures?

_____ [1 mark]

c Which two towns had a difference of 30 degrees between the maximum and minimum temperatures?

_____ [1 mark]

d Which town had the greatest difference between the maximum and minimum temperatures?

_____ [1 mark]

2 a The number line has the value –2.2 marked on it.

–2.2

-3   -2   -1   0   1   2   3

Mark the following values on the number line.

i –1.4 **ii** 1.7 **iii** –0.3 [2 marks]

b Fill in the missing values on this number line.

–1.9   –1.8   –1.7

[1 mark]

c What number is halfway between:

i –4 and 8 **ii** –11 and –8? [1 mark each]

_____   _____

This page tests you on   • negative numbers   • the number line

1 Look at the following number cards.

$+8$ $+6$ $+3$ 0 -5 -6 -7

a What is the total of the numbers on all the cards?

_____ **[1 mark]**

b Which two cards will make this calculation true?

[ ] + [ ] = **0**

_____ and _____ **[1 mark]**

c i Which card would make the answer to this calculation as small as possible?

$+4$ − [ ] = [ ]

_____ **[1 mark]**

ii What is the smallest possible answer?

_____ **[1 mark]**

d i Which card would make the answer to this calculation as large as possible?

$+4$ − [ ] = [ ]

_____ **[1 mark]**

ii What is the largest possible answer?

_____ **[1 mark]**

2 In this magic square the numbers in each row, column and diagonal add up to the same total.

Fill in the missing numbers.

3		
−3		
−6	7	−7

[2 marks]

This page tests you on • addition and subtraction with negative numbers

More about number

1 Here are six number cards.

a Which **two** of the numbers are multiples of 3?

_____ [1 mark]

b Which **two** of the numbers are factors of 30?

_____ [1 mark]

c In this magic square the numbers in each row, column and diagonal add up to the same total. Use numbers from the cards above to complete the magic square.

		8
5	9	

[2 marks]

F

2 a Write down the factors of each number.

i 33

_____ [1 mark]

ii 18

_____ [1 mark]

b From this list of numbers:

84, 85, 86, 88, 89, 90

write down:

i a multiple of 3 _____ [1 mark]

ii a multiple of 5. _____ [1 mark]

c Counter A counts a beat every 3 seconds.

Counter B counts a beat every 4 seconds

Counter C counts a beat every 5 seconds.

They all start at the same time.

After how many seconds will all three counters next count a beat at the same time?

_____ [1 mark]

C

This page tests you on • multiples • factors

Primes and squares

D

1 Here are seven number cards.

$$6 \quad 9 \quad 10 \quad 11 \quad 13 \quad 15 \quad 16$$

a Which **two** of the numbers are prime numbers?

_____ **[1 mark]**

b Jen says that all prime numbers are odd.

Give an example to show that Jen is wrong.

_____ **[1 mark]**

c Which **two** of the numbers are square numbers?

_____ **[1 mark]**

d Ken says that all square numbers end in the digits 1, 4, 6 or 9.

Give an example to show that Ken is wrong.

_____ **[1 mark]**

e Which **two** of the cards will give a result that is a prime number in the calculation below?

$$6 \quad + \quad \boxed{} \quad = \quad \text{prime}$$

_____ **[2 marks]**

D

2 P is a prime number. S is a square number, Q is an odd number.

Look at each of these expressions and decide whether it is *always even*, *always odd* or can be *either odd or even*. Tick the correct box.

	Always even	Always odd	Either odd or even	
a $P + S$	☐	☐	☐	**[1 mark]**
b $P \times S$	☐	☐	☐	**[1 mark]**
c Q^2	☐	☐	☐	**[1 mark]**
d $P + Q$	☐	☐	☐	**[1 mark]**

This page tests you on • prime numbers • square numbers

Roots and powers

1 Here are six number cards.

$$\boxed{4} \quad \boxed{5} \quad \boxed{6} \quad \boxed{7} \quad \boxed{8} \quad \boxed{9}$$

a Use **three** of the cards to make this statement true.

$$\sqrt{\boxed{}\boxed{}} = \boxed{}$$

[1 mark]

b Use **three different** cards to make this statement true.

$$\sqrt{\boxed{}\boxed{}} = \boxed{}$$

[1 mark]

c Which is greater, $\sqrt{144}$ or 2^4?

_____ [1 mark]

d Write down the value of each number.

  i $\sqrt{169}$ **ii** 5^3

_____   _____ [1 mark each]

e Write down the value of each number.

  i $\sqrt[3]{64}$ **ii** 2^8

_____   _____ [1 mark each]

E

D

2 a Fill in the missing numbers.

			Units digit
4^1	=	4	4
4^2	=	16	6
4^3	=	——	——
4^4	=	——	——
4^5	=	——	——

[2 marks]

b What is the last digit of 4^{99}? Explain your answer.

_____ [1 mark]

c Which is greater, 5^6 or 6^5? Justify your answer.

_____ [1 mark]

This page tests you on   • square roots   • powers

Powers of 10

D

1 a What is the value of the digit **4** in the number 23.4?

_____ **[1 mark]**

b Write 10 000 in the form 10^n, where n is an integer.

_____ **[1 mark]**

c Write, in full, the number represented by 10^7.

_____ **[1 mark]**

d Fill in the missing numbers and powers.

1000		10	1	$\frac{1}{10}$		$\frac{1}{1000}$
10^3	10^2	$10^{\square}$	$10^{\square}$	10^{-1}	10^{-2}	$10^{\square}$

[1 mark]

e Write down the value of:

i 6^0 _____ **[1 mark]**

ii 5^1 _____ **[1 mark]**

D

2 a Work out:

i 3.7×10^2

_____ **[1 mark]**

ii 0.25×10^3

_____ **[1 mark]**

b Work out:

i $7.6 \div 10$

_____ **[1 mark]**

ii $0.65 \div 10^2$

_____ **[1 mark]**

c Work out:

i $30\ 000 \times 400$

_____ **[1 mark]**

ii 600^2

_____ **[1 mark]**

d Work out:

i $90\ 000 \div 30$

_____ **[1 mark]**

ii $30\ 000 \div 60$

_____ **[1 mark]**

This page tests you on
- powers of 10
- multiplying and dividing by powers of 10
- multiplying and dividing multiples of powers of 10

Prime factors

---

1 a What number is represented by $2 \times 3^2 \times 5$?

_____ **[1 mark]**

b Write 70 as the product of its prime factors.

_____ **[1 mark]**

c Write 48 as the product of its prime factors.

_____ **[1 mark]**

d i Complete the prime factor tree to find the prime factors of 900.

100

10

900

[2 marks]

ii Write 900 as the product of its prime factors, in index form.

_____ **[1 mark]**

---

2 a i You are given that $3x^2 = 75$.

What is the value of x?

$x =$ _____ **[1 mark]**

ii Write 150 as the product of its prime factors.

$x =$ _____ **[1 mark]**

b i You are given that $2x^3 = 54$.

What is the value of x?

$x =$ _____ **[1 mark]**

ii Write 216 as the product of its prime factors.

_____ **[1 mark]**

This page tests you on • prime factors

LCM and HCF

C

1 a Write 24 as the product of its prime factors.

_____ **[1 mark]**

b Write 60 as the product of its prime factors.

_____ **[1 mark]**

c What is the lowest common multiple of 24 and 60?

_____ **[1 mark]**

d What is the highest common factor of 24 and 60?

_____ **[1 mark]**

e In prime factor form, the number $P = 2^4 \times 3^2 \times 5$ and the number $Q = 2^2 \times 3 \times 5^2$.

i What is the lowest common multiple of P and Q?

Give your answer in index form.

_____ **[1 mark]**

ii What is the highest common factor of P and Q?

Give your answer in index form.

_____ **[1 mark]**

C

2 a You are told that p and q are prime numbers.

$p^2q^2 = 36$

What are the values of p and q?

$p = $ _____

$q = $ _____ **[2 marks]**

b Write 360 as the product of its prime factors.

_____ **[1 mark]**

c You are told that a and b are prime numbers.

$ab^2 = 98$

What are the values of a and b?

$a = $ _____

$b = $ _____ **[2 marks]**

d Write 196 as the product of its prime factors.

_____ **[1 mark]**

This page tests you on • lowest common multiple • highest common factor

Powers

1 a Write $4^3 \times 4^5$ as a single power of 4.

_____ **[1 mark]**

b Write $6^5 \div 6^2$ as a single power of 6.

_____ **[1 mark]**

c i If $3^n = 81$, what is the value of n?

_____ **[1 mark]**

ii If $3^m = 27$, what is the value of m?

_____ **[1 mark]**

d Write down the actual value of $7^9 \div 7^7$.

_____ **[1 mark]**

e Write down the actual value of $10^4 \times 10^4$.

_____ **[1 mark]**

f Write down the actual value of $2^2 \times 5^2 \times 2^4 \times 5^4$.

_____ **[1 mark]**

2 a Write $x^5 \times x^2$ as a single power of x.

_____ **[1 mark]**

b Write $x^8 \div x^4$ as a single power of x.

_____ **[1 mark]**

c $2^3 \times 3^3 = 6^3$.

Which of the following expressions is the same as $a^n \times b^n$?

$(a + b)^n$ $\qquad$ ab^{2n} $\qquad$ $(ab)^n$

_____ **[1 mark]**

d $8^3 \div 2^3 = 4^3$.

Which of the following expressions is the same as $a^n \div b^n$?

$(a \div b)^n$ $\qquad$ $a \div b^n$ $\qquad$ $a^n - b^n$

_____ **[1 mark]**

This page tests you on
- multiplying powers • dividing powers
- multiplying and dividing powers with letters

Number skills

F

1 a Arne uses the column method to work out 37×48. This is his working.

```
          3    7
    ×     4    8
    2    9₅    6
    1    4₂    8
    4    4    4
    1    1
```

Arne has made a mistake.

i What mistake has Arne made?

_____ **[1 mark]**

ii Work out the correct answer to 37×48.

_____ **[1 mark]**

b Berne is using the box method to work out 29×47. This is his working.

×	20	9
40	60	49
7	27	16

```
          6    0
          4    9
          2    7
    +     1    6
    1    5    2
```

Berne has made a mistake.

i What mistake has Berne made?

_____ **[1 mark]**

ii Work out the correct answer to 29×47.

_____ **[2 marks]**

F

2 a There are 144 plasters in a box. How many plasters will there be in 24 boxes?

_____ **[2 marks]**

b Each box of plasters costs 98p. How much will 24 boxes of plasters cost?

Give your answer in pounds and pence.

_____ **[2 marks]**

This page tests you on • long multiplication

131

1 a Write down the answers to these.

F

    i $1 \times 29 =$ _____

    ii $2 \times 29 =$ _____

    iii $10 \times 29 =$ _____

    iv $20 \times 29 =$ _____ **[2 marks]**

b Using the values above, or otherwise, complete this division by the chunking method.

$1508 \div 29$
```
      1  5  0  8
   -     5  8  0
```

[2 marks]

c i Write down the answer to $1520 \div 29$.

_____ **[1 mark]**

   ii Write down the answer to $1508 \div 58$.

_____ **[1 mark]**

2 a A widget machine produces 912 widgets per hour.

F

How many widgets does it produce during a 16-hour shift?

_____ **[2 marks]**

b The widgets are packed in boxes of 24.

How many boxes will be needed to pack 912 widgets?

_____ **[2 marks]**

This page tests you on   • long division

F

1 a The school canteen has 34 tables.

Each table can seat up to 14 students.

What is the maximum number of people who could use the canteen at one time?

_____ **[2 marks]**

b The head wants to expand the canteen so it can cater for 600 students.

How many extra tables will be needed?

_____ **[2 marks]**

F

2 a This table shows the column headings for the number 23.4789.

10	1	•	$\frac{1}{10}$	$\frac{1}{100}$	$\frac{1}{1000}$	$\frac{1}{10\,000}$
2	3	•	4	7	8	9

i The number 23.4789 is multiplied by 100.

What will be the place value of the digit 4 in the **answer** to 23.4789 × 100?

_____ **[1 mark]**

ii The number 23.4789 is divided by 10.

What will be the place value of the digit 7 in the **answer** to 23.4789 ÷ 10?

_____ **[1 mark]**

b Round the number 23.4789 to:

i 1 decimal place

_____ **[1 mark]**

ii 2 decimal places

_____ **[1 mark]**

iii 3 decimal places.

_____ **[1 mark]**

This page tests you on • real-life problems • decimal places

E

1 a Complete the shopping bill.

3 jars of jam at £1.28 per jar	
2 kg of apples at £2.15 per kg	
5 doughnuts at 22p each	
Total	**[4 marks]**

b Work these out.

i 3×2.6

_____ **[1 mark]**

ii 2.4×2.6

_____ **[2 marks]**

c A dividend is a repayment made every few months on the amount spent.
The Co-op pays a dividend of 2.6 pence for every £1 spent.

i In three months, Derek spends £240.

How much dividend will he get?

_____ **[1 mark]**

ii Doreen received a dividend of £7.80.

How much did she spend to get this dividend?

_____ **[1 mark]**

F

2 a i Work out 6×2.9.

_____ **[1 mark]**

ii Work out 4.6×2.9.

_____ **[2 marks]**

b What is the cost of 2.9 kg of coffee beans at £4.60 per kilogram?

_____ **[1 mark]**

This page tests you on
- adding and subtracting decimals
- multiplying and dividing decimals by single-digit numbers
- long multiplication with decimals
- multiplying decimals

More fractions

D-C

1 a Work out $\frac{3}{4} + \frac{2}{5}$.

Give your answer as a mixed number.

_____ **[2 marks]**

b Work out $3\frac{2}{3} - 1\frac{4}{5}$.

Give your answer as a mixed number.

_____ **[2 marks]**

c On an aeroplane, two-fifths of the passengers were British, one-quarter were German, one-sixth were American and the rest were French.

What fraction of the passengers are French?

_____ **[2 marks]**

D-C

2 a Work out $2\frac{1}{2} \times 1\frac{2}{5}$.

_____ **[2 marks]**

b Work out $3\frac{3}{10} \div 2\frac{2}{5}$.

Give your answer as a mixed number.

_____ **[2 marks]**

c Work out the area of this triangle.

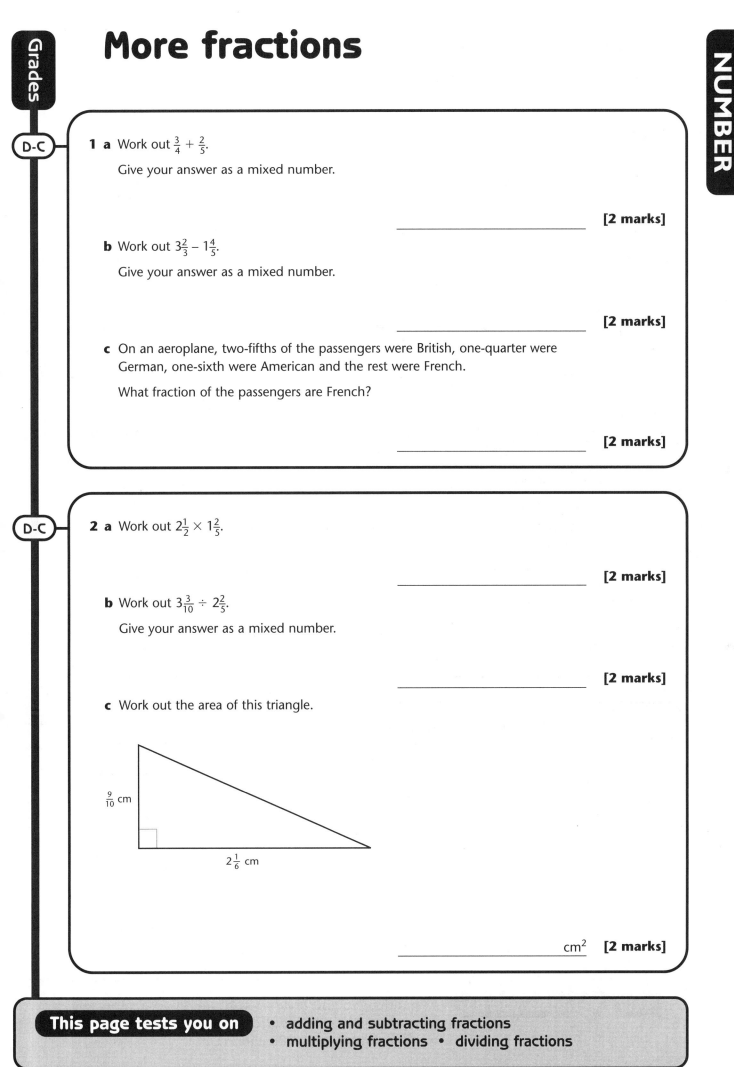

$\frac{9}{10}$ cm

$2\frac{1}{6}$ cm

_____ cm^2 **[2 marks]**

This page tests you on • adding and subtracting fractions
• multiplying fractions • dividing fractions

More number

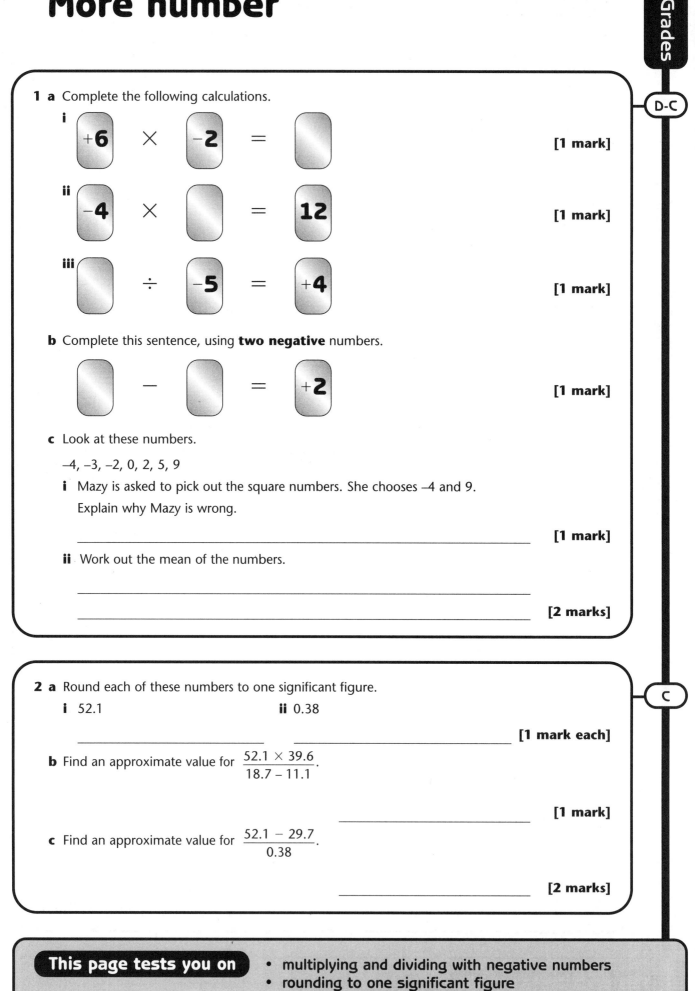

1 a Complete the following calculations.

    i $+6$ × -2 = ☐   **[1 mark]**

    ii -4 × ☐ = 12 **[1 mark]**

    iii   ☐ ÷ -5 = $+4$ **[1 mark]**

b Complete this sentence, using **two negative** numbers.

     ☐ − ☐ = $+2$ **[1 mark]**

c Look at these numbers.

    −4, −3, −2, 0, 2, 5, 9

    i Mazy is asked to pick out the square numbers. She chooses −4 and 9.

    Explain why Mazy is wrong.

    _____ **[1 mark]**

    ii Work out the mean of the numbers.

    _____

    _____ **[2 marks]**

D–C

2 a Round each of these numbers to one significant figure.

    i 52.1   **ii** 0.38

    _____   _____ **[1 mark each]**

b Find an approximate value for $\dfrac{52.1 \times 39.6}{18.7 - 11.1}$.

    _____ **[1 mark]**

c Find an approximate value for $\dfrac{52.1 - 29.7}{0.38}$.

    _____ **[2 marks]**

C

This page tests you on
- multiplying and dividing with negative numbers
- rounding to one significant figure
- approximation of calculations

Ratio

C

1 The angles of a quadrilateral are in the ratio $2:3:5:8$.

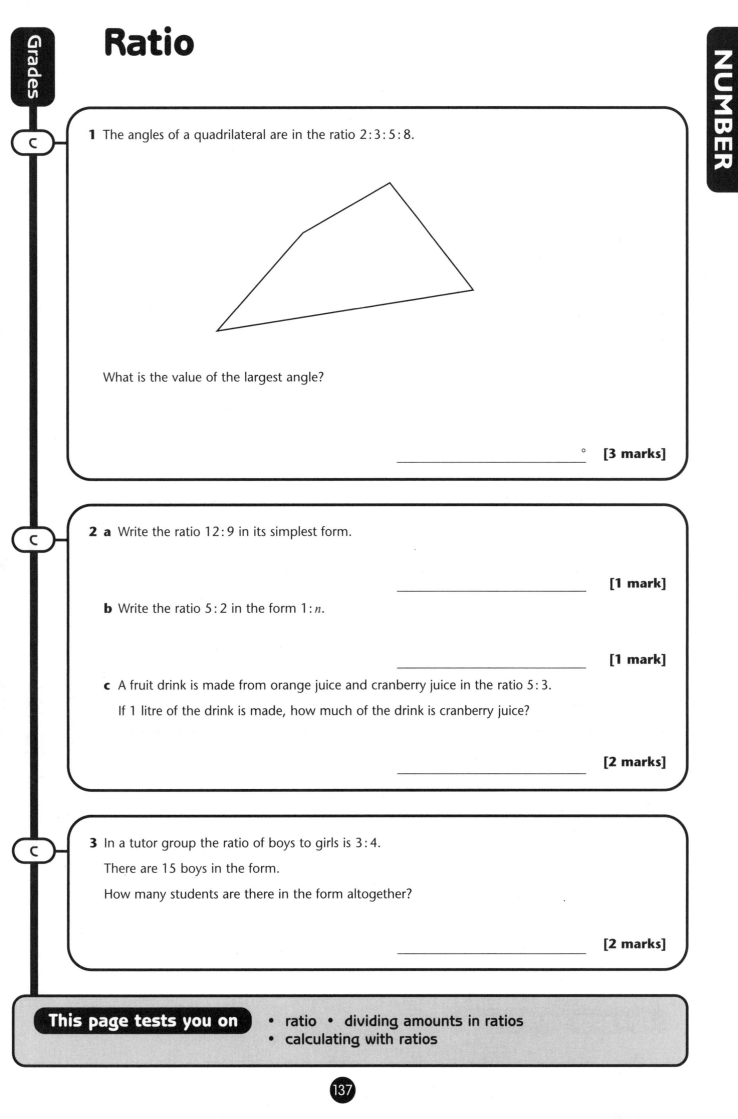

What is the value of the largest angle?

_____ ° **[3 marks]**

C

2 a Write the ratio $12:9$ in its simplest form.

_____ **[1 mark]**

b Write the ratio $5:2$ in the form $1:n$.

_____ **[1 mark]**

c A fruit drink is made from orange juice and cranberry juice in the ratio $5:3$.

If 1 litre of the drink is made, how much of the drink is cranberry juice?

_____ **[2 marks]**

C

3 In a tutor group the ratio of boys to girls is $3:4$.

There are 15 boys in the form.

How many students are there in the form altogether?

_____ **[2 marks]**

This page tests you on • ratio • dividing amounts in ratios
• calculating with ratios

Speed and proportion

1 A ferry covers the 72 kilometres between Holyhead and Dublin in $2\frac{1}{4}$ hours.

D-C

a What is the average speed of the ferry?

State the units of your answer.

_____ **[3 marks]**

b For the first 15 minutes and the last 15 minutes, the Ferry is manoeuvring in and out of the ports. During this time the average speed is 18 km per hour.

What is the average speed of the ferry during the rest of the journey?

_____ **[2 marks]**

2 A car uses 50 litres of petrol in driving 275 miles.

D

a How much petrol will the car use in driving 165 miles?

_____ **[2 marks]**

b How many miles can the car drive on 26 litres of petrol?

_____ **[1 mark]**

3 Nutty Flake cereal is sold in two sizes.

D

The handy size contains 600 g and costs £1.55.

The large size contains 800 g and costs £2.20.

Which size is the better value?

_____ **[2 marks]**

This page tests you on
- speed, time and distance
- direct proportion problems • best buys

Percentages

1 a Complete this table of equivalent fractions, decimals and percentages.

Decimal	Fraction	Percentage
0.35		
	$\frac{4}{5}$	
		90%

[3 marks]

b i What fraction of this diagram is shaded?

Give your answer in its simplest form.

_____ [1 mark]

ii What percentage is **not** shaded?

_____ [1 mark]

c Write these values in order of size, smallest first.

0.6 $\quad \frac{11}{20} \quad$ 57% $\quad \frac{14}{25}$

_____ [2 marks]

2 a A car cost £6000 new.

It depreciated in value by 12% in the first year.

It then depreciated in value by 10% in the second year.

Which of these calculations shows the value of the car after two years?

Circle the correct answer.

$6000 \times 0.78 \qquad 6000 \times 0.88 \times 0.9 \qquad 6000 \times 88 \times 90 \qquad 6000 - 2200$ [1 mark]

b VAT is charged at $17\frac{1}{2}$%.

A quick way to work out the VAT on an item is to work out 10%, then divide this by 2 to get 5%, then divide this by 2 to get $2\frac{1}{2}$%. Add these values all together.

Use this method to work out the VAT on an item costing £68.

_____ [2 marks]

This page tests you on
• equivalent percentages, fractions and decimals
• the percentage multiplier

1 A washing machine normally costs £350.

Its price is reduced by 15% in a sale.

a What is 15% of £350?

£ _____ [2 marks]

b What is the sale price of the washing machine?

£ _____ [1 mark]

E

2 a A computer costs £700, not including VAT.

VAT is charged at $17\frac{1}{2}$%.

What is the cost of the computer when VAT is added?

£ _____ [2 marks]

b The price of a printer is reduced by 12% in a sale.

The original price of the printer was £250.

What is the price of the printer in the sale?

£ _____ [2 marks]

D

3 In the first week it was operational, a new bus route carried a total of 2250 people.

In the second week it carried 2655 people.

What is the percentage increase in the number of passengers from the first week to the second?

_____ [2 marks]

C

This page tests you on
- calculating a percentage of a quantity
- percentage increase or decrease
- one quantity as a percentage of another

Number checklist

I can...

☐ recall the times tables up to 10×10

☐ use BODMAS to do calculations in the correct order

☐ identify the place value of digits in whole numbers

☐ round numbers to the nearest 10 or 100

☐ add and subtract numbers with up to four digits without a calculator

☐ multiply numbers by a single-digit number

☐ state what fraction of a shape is shaded

☐ shade in a fraction of a shape

☐ add and subtract simple fractions with the same denominator

☐ recognise equivalent fractions

☐ cancel a fraction

☐ change top-heavy fractions into mixed numbers and vice versa

☐ find a fraction of an integer

☐ recognise the multiples of the first 10 whole numbers

☐ recognise square numbers up to 100

☐ find equivalent fractions, decimals and percentages

☐ know that a number on the left on a number line is smaller than a number on the right

☐ read negative numbers on scales such as thermometers

You are working at ⬭ Grade G ⬭ level.

☐ divide numbers by a single-digit number

☐ put fractions in order of size

☐ add fractions with different denominators

☐ solve fraction problems expressed in words

☐ compare fractions of quantities

☐ find factors of numbers less than 100

☐ add and subtract with negative numbers

☐ write down the squares of numbers up to 15×15

☐ write down the cubes of 1, 2, 3, 4, 5 and 10

☐ use a calculator to find square roots

☐ do long multiplication

☐ do long division

☐ solve real-life problems involving multiplication and division

☐ round decimal numbers to one, two or three decimal places

☐ find percentages of a quantity

☐ change mixed numbers into top-heavy fractions

You are working at ⬭ Grade F ⬭ level.

☐ multiply fractions

☐ add and subtract mixed numbers

☐ calculate powers of numbers

☐ recognise prime numbers under 100

☐ use the four rules with decimals

☐ change decimals to fractions

☐ change fractions to decimals

☐ simplify a ratio

☐ find a percentage of any quantity

You are working at (Grade E) level.

☐ work out one quantity as a fraction of another

☐ solve problems using negative numbers

☐ multiply and divide by powers of 10

☐ multiply together numbers that are multiples of powers of 10

☐ round numbers to one significant figure

☐ estimate the answer to a calculation

☐ order lists of numbers containing decimals, fractions and percentages

☐ multiply and divide fractions

☐ calculate with speed, distance and time

☐ compare prices to find 'best buys'

☐ find the new value after a percentage increase or decrease

☐ find one quantity as a percentage of another

☐ solve problems involving simple negative numbers

☐ multiply and divide fractions

You are working at (Grade D) level.

☐ work out a reciprocal

☐ recognise and work out terminating and recurring decimals

☐ write a number as a product of prime factors

☐ use the index laws to simplify calculations and expressions

☐ multiply and divide with negative numbers

☐ multiply and divide with mixed numbers

☐ find a percentage increase

☐ work out the LCM and HCF of two numbers

☐ solve problems using ratio in appropriate situations

You are working at (Grade C) level.

Perimeter and area

F

1 Here is a rectangle.

a Find the perimeter of the rectangle.

State the units of your answer.

_____

b Find the area of the rectangle.

State the units of your answer.

59 cm

78 cm

[2 marks]

[2 marks]

G

2 a The map of an island is drawn on the centimeter-square grid.

The scale is 1 cm represents 1 km.

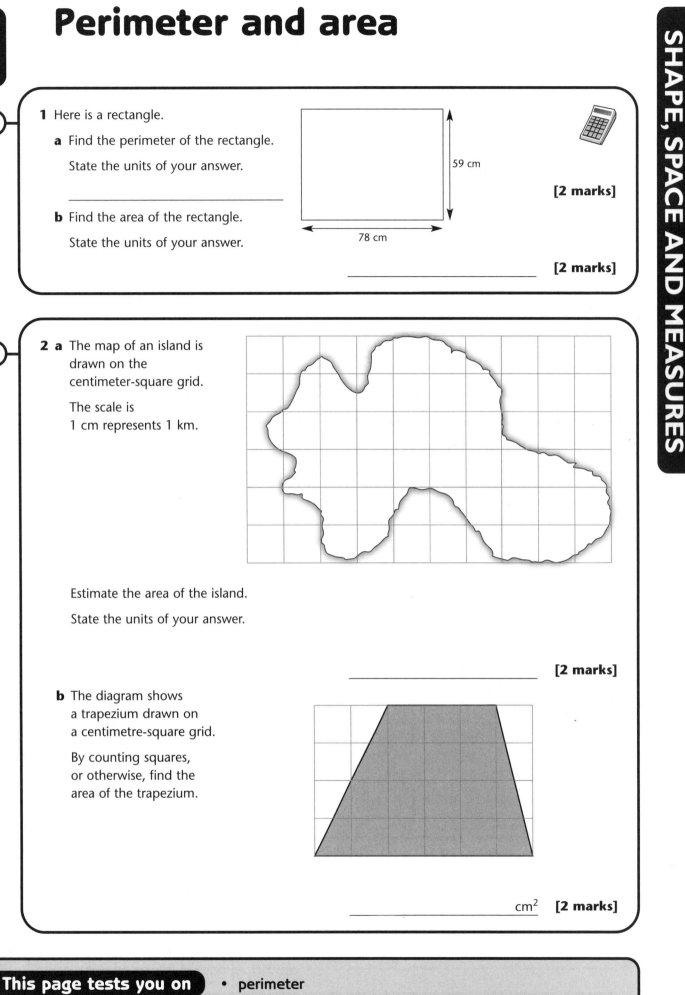

Estimate the area of the island.

State the units of your answer.

_____ **[2 marks]**

b The diagram shows a trapezium drawn on a centimetre-square grid.

By counting squares, or otherwise, find the area of the trapezium.

_____ cm² **[2 marks]**

This page tests you on
- perimeter
- area of irregular shapes (counting squares)
- area of a rectangle

1 Calculate the area of this shape.

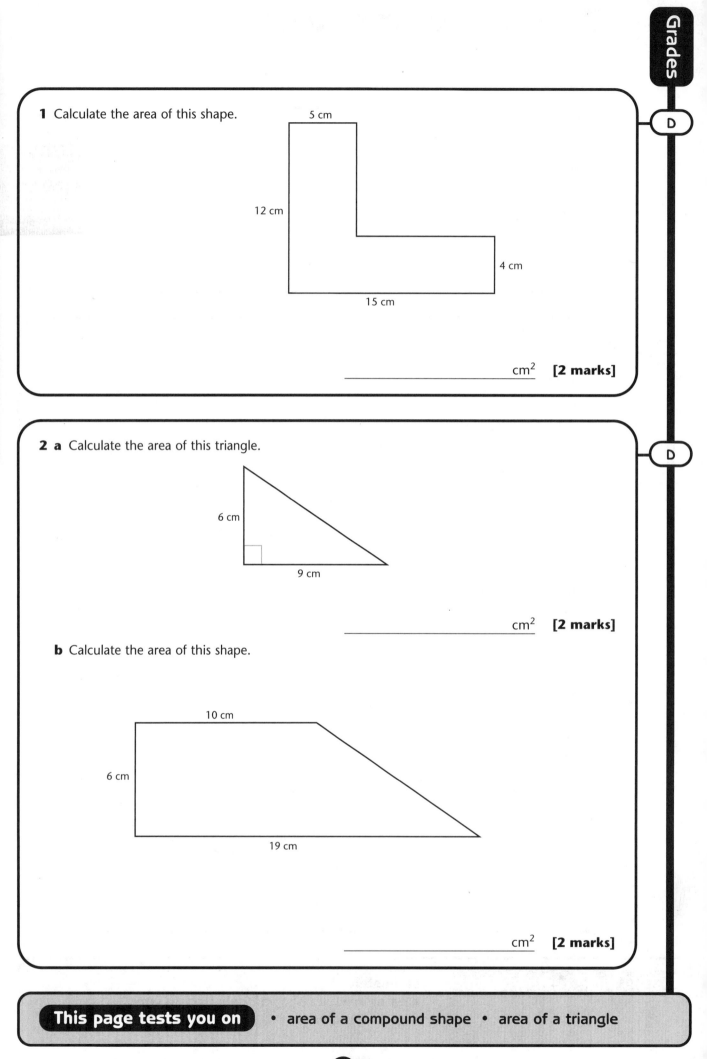

D

5 cm

12 cm

4 cm

15 cm

_____ cm² **[2 marks]**

2 a Calculate the area of this triangle.

D

6 cm

9 cm

_____ cm² **[2 marks]**

b Calculate the area of this shape.

10 cm

6 cm

19 cm

_____ cm² **[2 marks]**

This page tests you on   • area of a compound shape   • area of a triangle

Grades

D

1 a The parallelogram is drawn on a centimetre-square grid.

Find the area of the parallelogram.

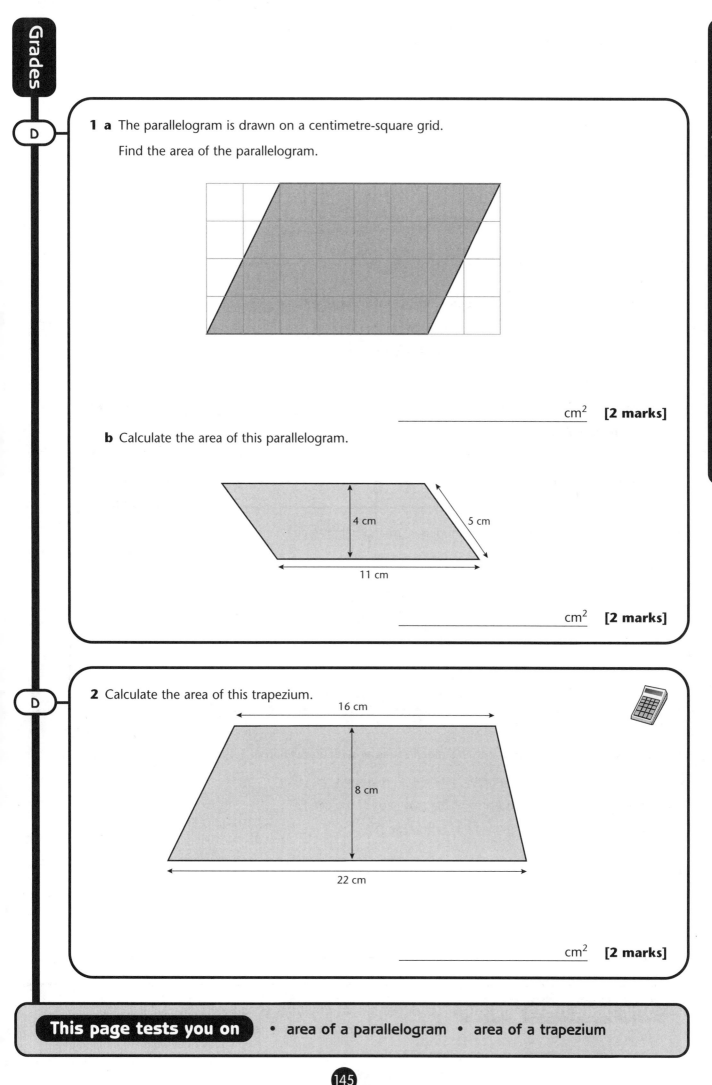

_____ cm² **[2 marks]**

b Calculate the area of this parallelogram.

4 cm

5 cm

11 cm

_____ cm² **[2 marks]**

D

2 Calculate the area of this trapezium.

16 cm

8 cm

22 cm

_____ cm² **[2 marks]**

This page tests you on • area of a parallelogram • area of a trapezium

Dimensional analysis

1 A cuboid has sides of length x, y and z cm.

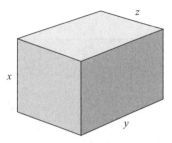

One of the following formulae represents the total lengths of the edges (L).

One of them represents the total area of the faces (A).

One of them represents the total volume (V).

Indicate which is which.

- $2xz + 2yz + 2xy$ represents the total _____ **[1 mark]**
- xyz represents the total _____ **[1 mark]**
- $4x + 4y + 4z$ represents the total _____ **[1 mark]**

2 A can of beans is a cylinder with a radius r cm and height h cm.

1 cm

The can has a label around it that is glued together with an overlap of 1 centimetre.

One of these formulae represents the perimeter of the label (P).

One of them represents the area of the two ends of the can (A).

One of them represents the volume of the can (V).

Indicate which is which.

- $2\pi r^2$ represents the _____ **[1 mark]**
- $4\pi r + 2h + 2$ represents the _____ **[1 mark]**
- $\pi r^2 h$ represents the _____ **[1 mark]**

This page tests you on • dimensional analysis • dimensions of length
• dimensions of area • dimensions of volume

Symmetry

G

1 EXAMS

a Which of the letters above have line symmetry?

_____ **[2 marks]**

b Which of the letters above have rotational symmetry of order 2?

_____ **[1 mark]**

c

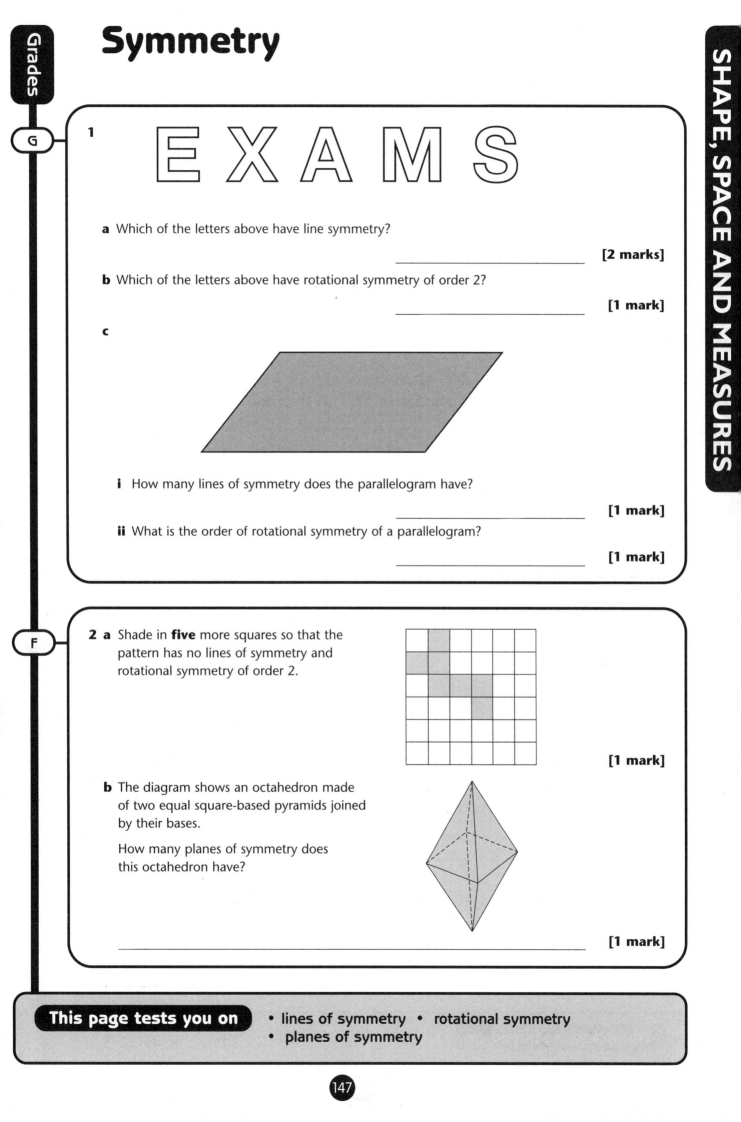

i How many lines of symmetry does the parallelogram have?

_____ **[1 mark]**

ii What is the order of rotational symmetry of a parallelogram?

_____ **[1 mark]**

F

2 a Shade in **five** more squares so that the pattern has no lines of symmetry and rotational symmetry of order 2.

[1 mark]

b The diagram shows an octahedron made of two equal square-based pyramids joined by their bases.

How many planes of symmetry does this octahedron have?

_____ **[1 mark]**

This page tests you on • lines of symmetry • rotational symmetry
• planes of symmetry

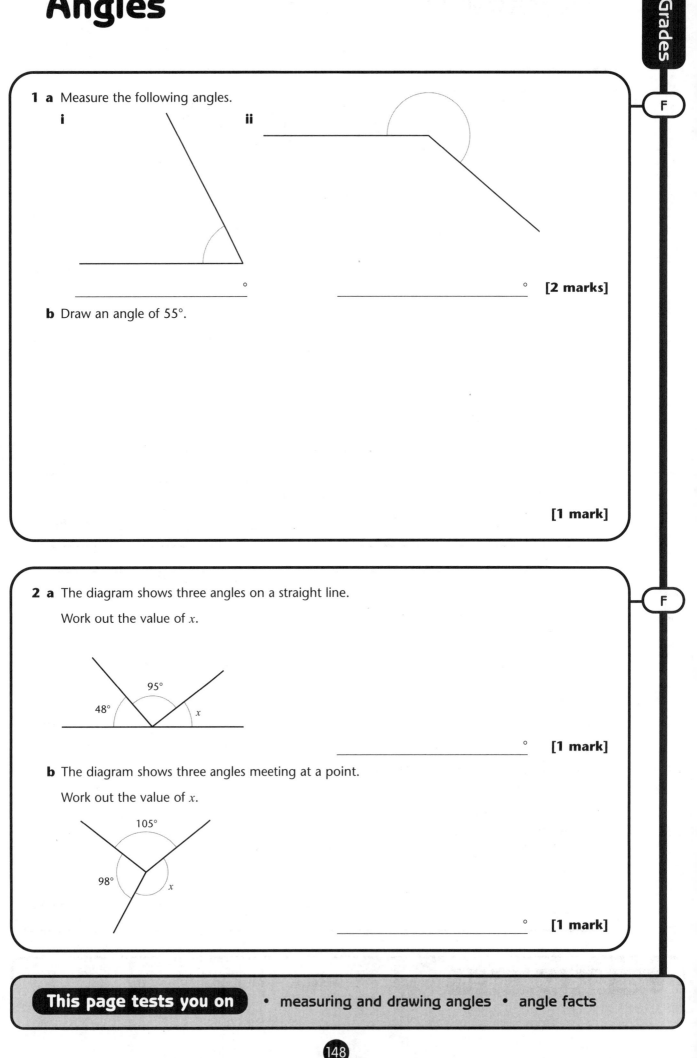

Angles

1 a Measure the following angles.

i

ii

_____ °   _____ ° **[2 marks]**

b Draw an angle of 55°.

[1 mark]

2 a The diagram shows three angles on a straight line.

Work out the value of x.

95°

48° x

_____ ° **[1 mark]**

b The diagram shows three angles meeting at a point.

Work out the value of x.

105°

98° x

_____ ° **[1 mark]**

This page tests you on   • measuring and drawing angles   • angle facts

SHAPE, SPACE AND MEASURES

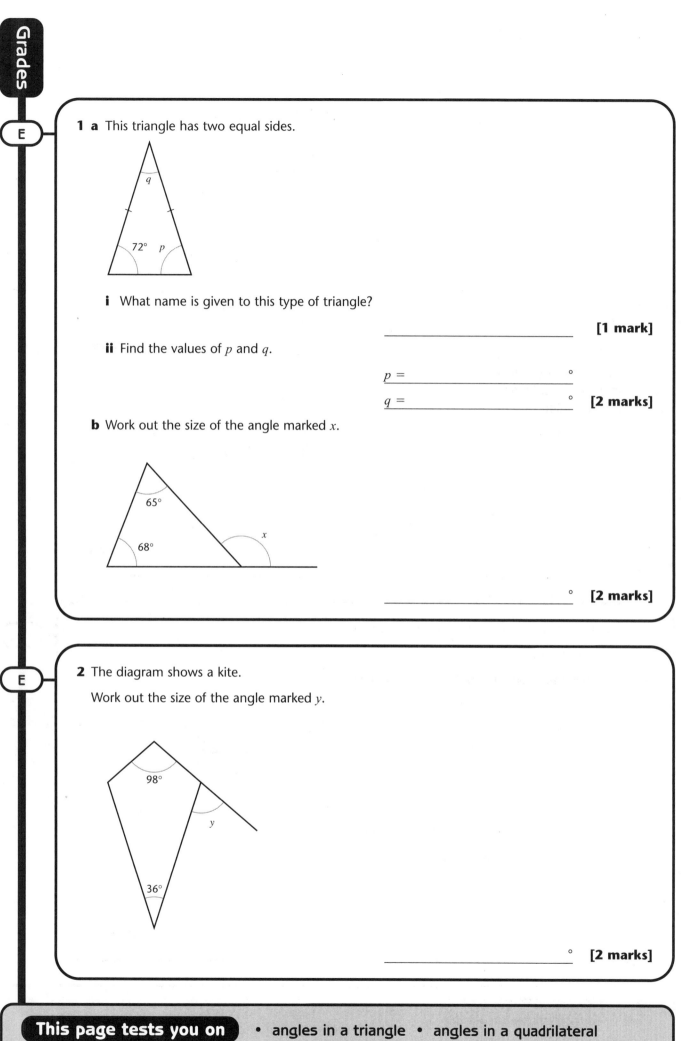

Grades

E

1 a This triangle has two equal sides.

    i What name is given to this type of triangle?

    _____ **[1 mark]**

    ii Find the values of p and q.

    $p =$ _____ °

    $q =$ _____ ° **[2 marks]**

b Work out the size of the angle marked x.

    65°

    68°

    x

_____ ° **[2 marks]**

E

2 The diagram shows a kite.

Work out the size of the angle marked y.

    98°

    y

    36°

_____ ° **[2 marks]**

This page tests you on   • angles in a triangle   • angles in a quadrilateral

Polygons

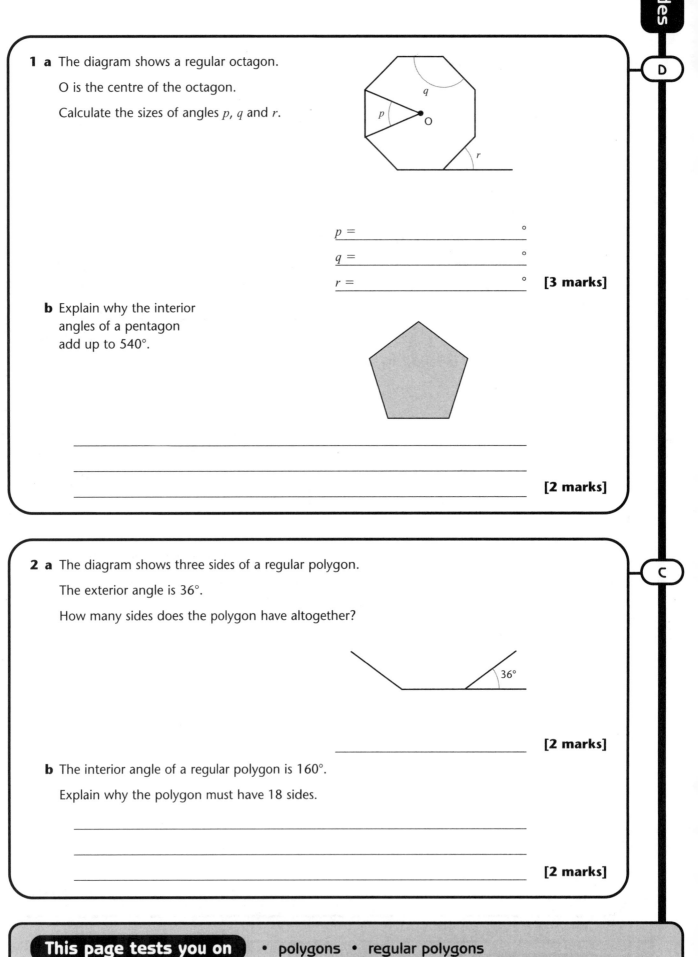

1 a The diagram shows a regular octagon.

O is the centre of the octagon.

Calculate the sizes of angles p, q and r.

$p =$ _____ °

$q =$ _____ °

$r =$ _____ ° **[3 marks]**

D

b Explain why the interior angles of a pentagon add up to 540°.

_____

_____

_____ **[2 marks]**

2 a The diagram shows three sides of a regular polygon.

The exterior angle is 36°.

How many sides does the polygon have altogether?

36°

_____ **[2 marks]**

C

b The interior angle of a regular polygon is 160°.

Explain why the polygon must have 18 sides.

_____

_____

_____ **[2 marks]**

This page tests you on • polygons • regular polygons
• interior and exterior angles in a regular polygon

Parallel lines and angles

C

1 In the diagram, QR is parallel to LM.

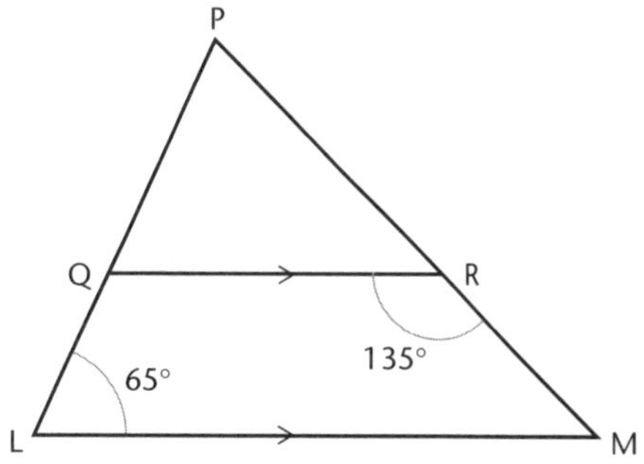

a Write down the size of angle RML.

Give a reason for your answer.

∠RML = _____ ° **[1 mark]**

Reason: _____ **[1 mark]**

b Write down the size of angle PQR.

Give a reason for your answer.

∠PQR = _____ ° **[1 mark]**

Reason: _____ **[1 mark]**

c Work out the size of angle QPR.

∠QPR = _____ ° **[1 mark]**

C

2 The lines AB and CD are parallel.

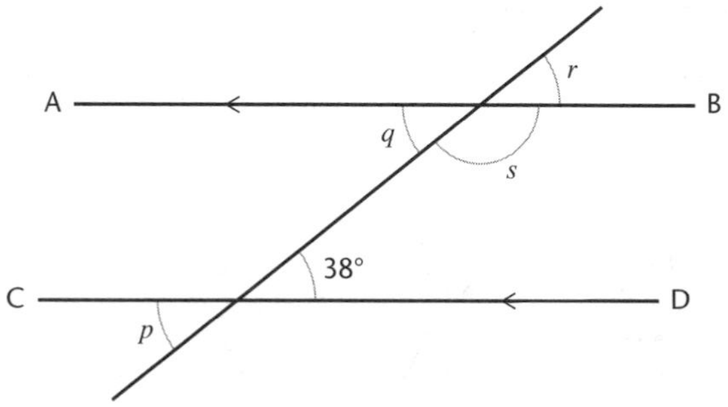

Write down the sizes of angles p, q, r and s, in each case give the reason in relation to the given angle of 38°.

p = _____ ° because it is _____ to the given angle of 38°. **[2 marks]**

q = _____ ° because it is _____ to the given angle of 38°. **[2 marks]**

r = _____ ° because it is _____ to the given angle of 38°. **[2 marks]**

s = _____ ° because it is _____ to the given angle of 38°. **[2 marks]**

This page tests you on   • two parallel lines and a transversal • alternate angles
  • corresponding angles • opposite angles
  • interior angles

Quadrilaterals

1 a Jonathan is describing a quadrilateral.

What quadrilateral is he describing?

D

It has no lines
of symmetry and has
rotational symmetry
of order 2.

The diagonals
cross at right angles
and all the sides
are equal.

_____ [1 mark]

b Marie is describing another quadrilateral.

i Write down the name of a quadrilateral
that Marie could be describing.

_____ [1 mark]

ii Write down the name of a different
quadrilateral that Marie could be
describing.

_____ [1 mark]

2 The diagram shows a kite, ABCD, attached
to a parallelogram, CDEF.

Angle BAD = 100°, angle CFE = 60°.

When the side of the kite is extended
it passes along the diagonal of the
parallelogram.

Use the properties of quadrilaterals
to work out the size of angle CED.

C

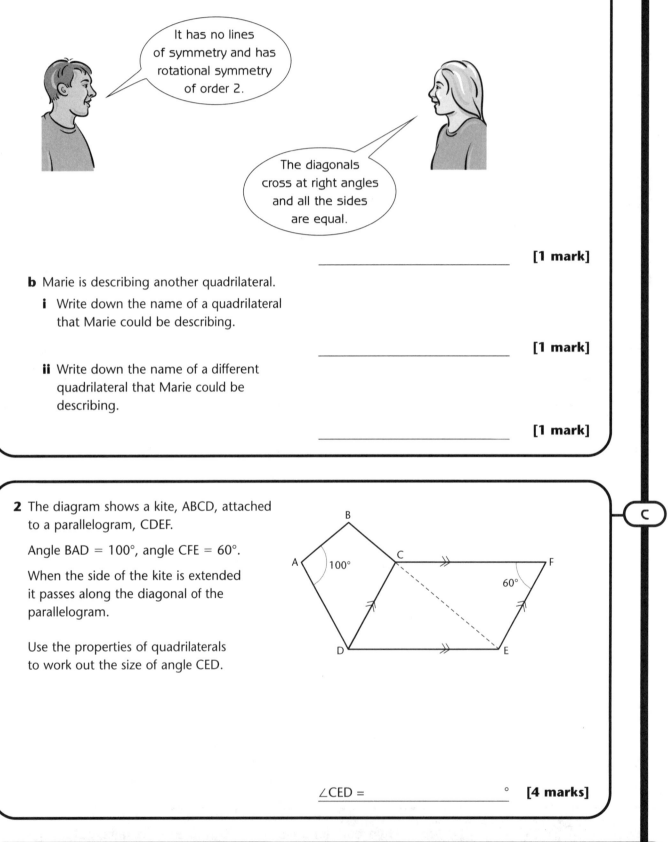

∠CED = _____ ° **[4 marks]**

Bearings

D

1 This map shows the positions of four towns, Althorp, Beeton, Cowton and Deepdale. The scale is 1 cm to 1 kilometre.

Beeton

N

Cowton

Althorp

Deepdale

Using a ruler and protractor, find the distance and bearing of:

a Beeton from Althorp _____ km at ____° **[2 marks]**

b Deepdale from Beeton _____ km at ____° **[2 marks]**

c Deepdale from Cowton _____ km at ____° **[2 marks]**

d Althorp from Deepdale _____ km at ____° **[2 marks]**

C

2 Brian walks in a perfect square.

He starts by walking north for 100 m and then turning right.

He then continues walking for 100 m then turning right, doing it twice more until he is back to his starting point.

Write down the four bearings of the directions in which he walks.

_____ **[4 marks]**

This page tests you on • bearings • measuring a bearing

Circles

1 From this list of words, fill in the missing words that describe parts of the circle on the diagram.

Chord   Tangent   Radius   Diameter

O is the centre of the circle.

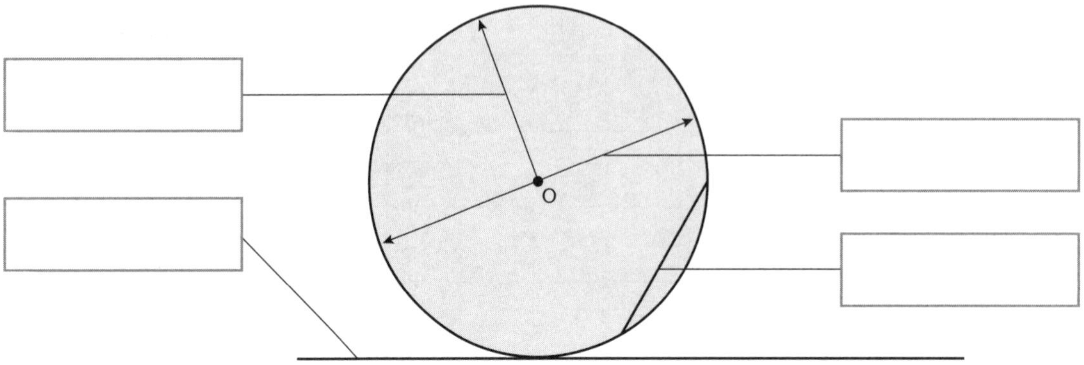

F

[4 marks]

2 a Work out the area of a circle of radius 12 cm.

Give your answer to 1 decimal place.

D

_____ cm² **[2 marks]**

b Work out the circumference of a circle of diameter 20 cm.

Give your answer to 1 decimal place.

_____ cm **[2 marks]**

3 Work out the area of a semicircle of diameter 20 cm.

Give your answer in terms of π.

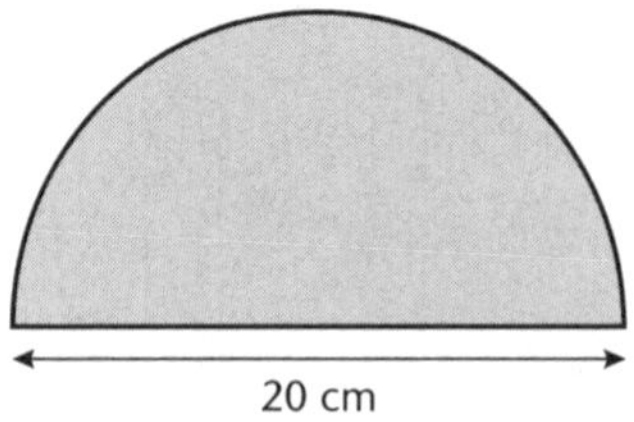

C

20 cm

_____ cm² **[2 marks]**

This page tests you on   • circles • circumference of a circle • area of a circle

Scales

G

1 a The thermometer shows the temperature outside Frank's local garage on a hot summer day. What temperature does the thermometer show?

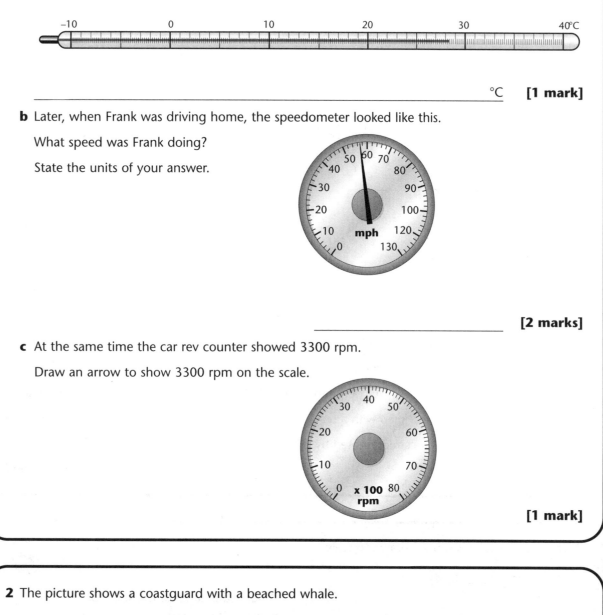

_____ °C   **[1 mark]**

b Later, when Frank was driving home, the speedometer looked like this.

What speed was Frank doing?

State the units of your answer.

_____ **[2 marks]**

c At the same time the car rev counter showed 3300 rpm.

Draw an arrow to show 3300 rpm on the scale.

[1 mark]

F

2 The picture shows a coastguard with a beached whale.

Estimate the length of the whale.

Give your answer in metres.

_____ m   **[2 marks]**

This page tests you on • scales • sensible estimates

Scales and drawing

1 The diagram shows a line AB and a point C, drawn on a centimetre-square grid.

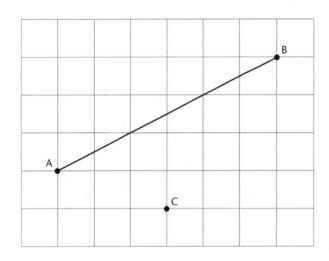

a Measure the length of the line AB, in centimetres. _____ cm **[1 mark]**

b Mark the midpoint of AB with a cross. **[1 mark]**

c Draw a line through the point C, perpendicular to the line AB. **[1 mark]**

d The diagram represents a map with a scale of 1 cm to 5 km. **[1 mark]**

Work out the real distance represented by BC.

_____ km **[2 marks]**

2 The net of a solid is shown, drawn to scale.

h

a What is the name of the solid for which this is the net?

_____ **[1 mark]**

b Measure the height of the triangle, *h*, shown on the net.

_____ cm **[1 mark]**

c By taking appropriate measurements, work out the surface area of the net.

_____ cm^2 **[3 marks]**

This page tests you on   • scale drawing • nets

3-D drawing

F

1 The diagram shows a prism, with a T-shaped cross-section, drawn on a one-centimetre isometric grid.

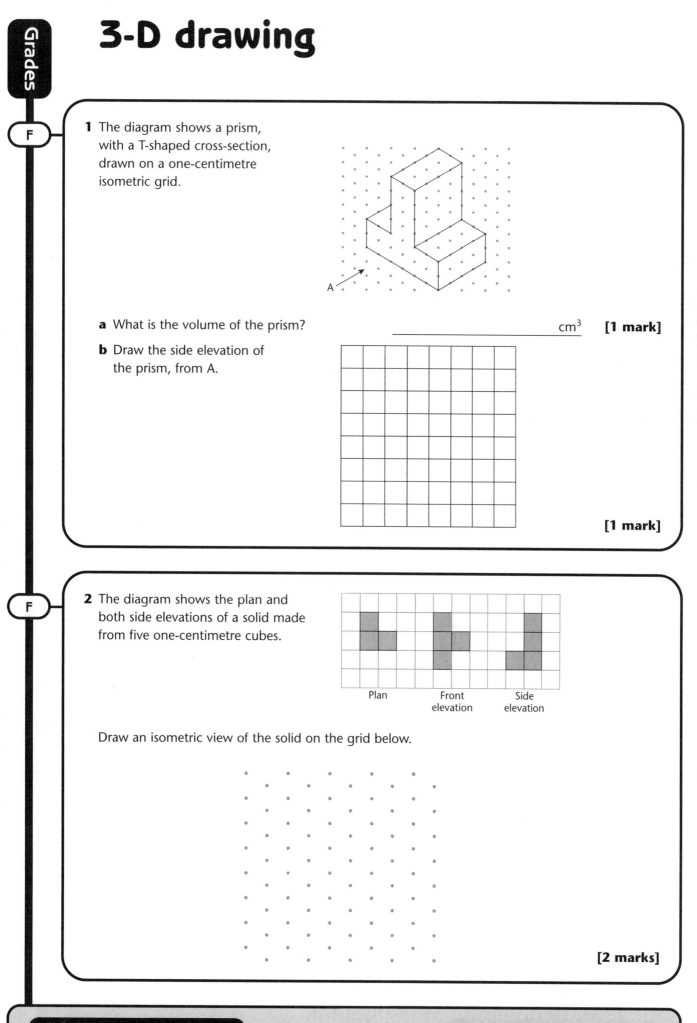

A

a What is the volume of the prism? _____ cm³ **[1 mark]**

b Draw the side elevation of the prism, from A.

[1 mark]

F

2 The diagram shows the plan and both side elevations of a solid made from five one-centimetre cubes.

Plan   Front elevation   Side elevation

Draw an isometric view of the solid on the grid below.

[2 marks]

This page tests you on   • isometric grids   • plans and elevations

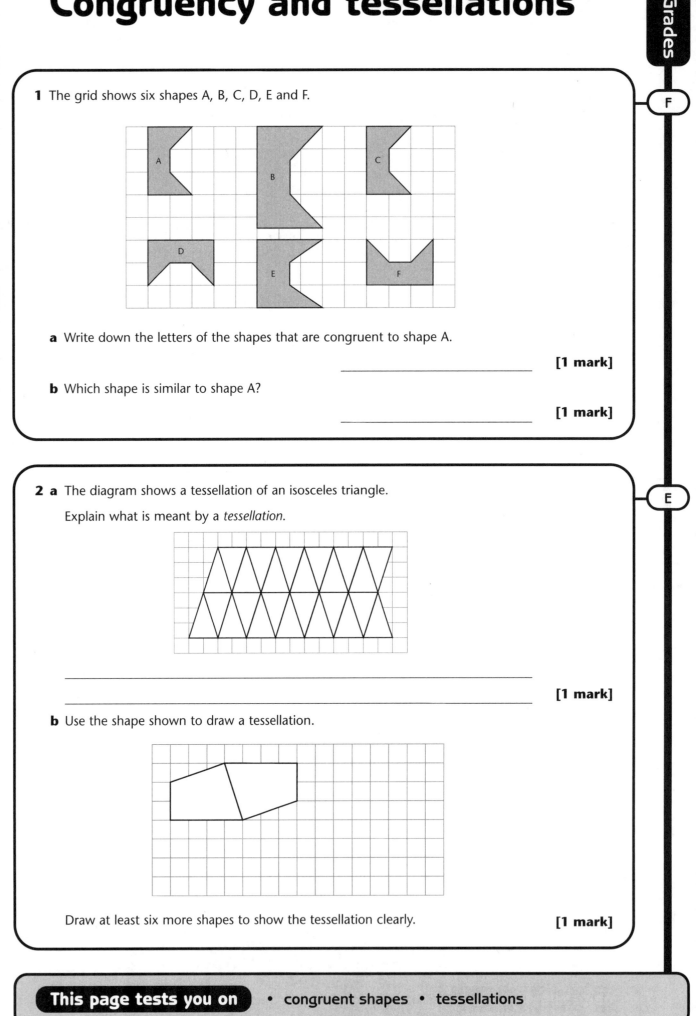

Congruency and tessellations

F

1 The grid shows six shapes A, B, C, D, E and F.

a Write down the letters of the shapes that are congruent to shape A.

_____ **[1 mark]**

b Which shape is similar to shape A?

_____ **[1 mark]**

E

2 a The diagram shows a tessellation of an isosceles triangle.

Explain what is meant by a *tessellation*.

_____

_____ **[1 mark]**

b Use the shape shown to draw a tessellation.

Draw at least six more shapes to show the tessellation clearly. **[1 mark]**

Transformations

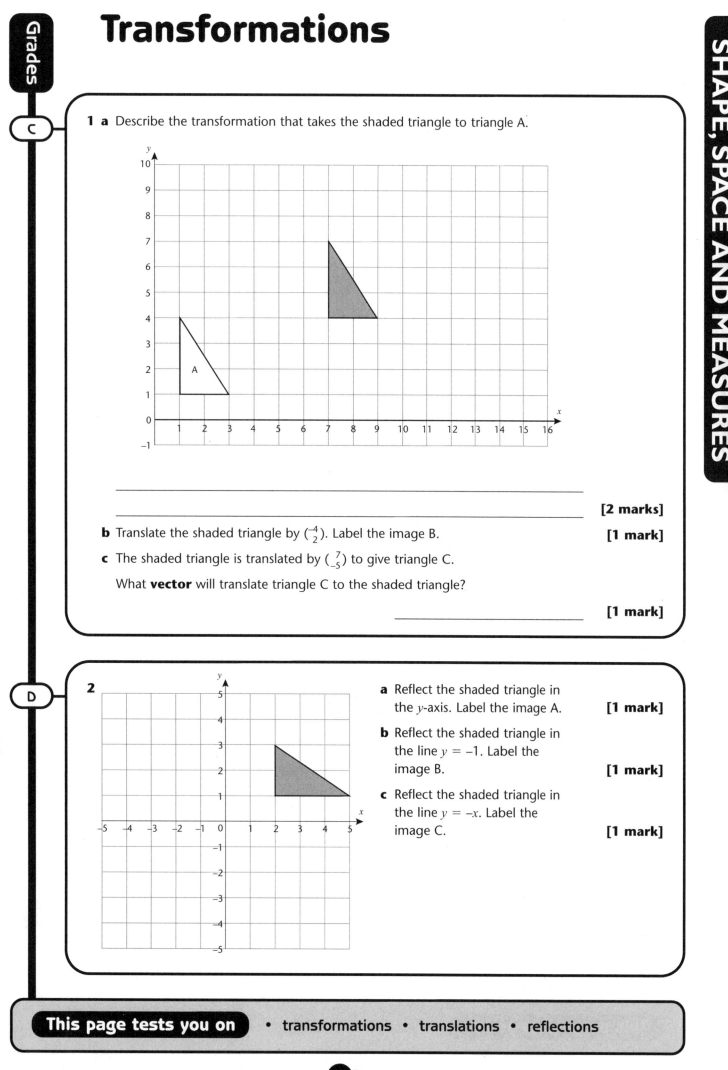

1 a Describe the transformation that takes the shaded triangle to triangle A.

C

_____

_____ **[2 marks]**

b Translate the shaded triangle by $\binom{-4}{2}$. Label the image B. **[1 mark]**

c The shaded triangle is translated by $\binom{7}{-5}$ to give triangle C.

What **vector** will translate triangle C to the shaded triangle?

_____ **[1 mark]**

D

2

a Reflect the shaded triangle in the y-axis. Label the image A. **[1 mark]**

b Reflect the shaded triangle in the line $y = -1$. Label the image B. **[1 mark]**

c Reflect the shaded triangle in the line $y = -x$. Label the image C. **[1 mark]**

This page tests you on • transformations • translations • reflections

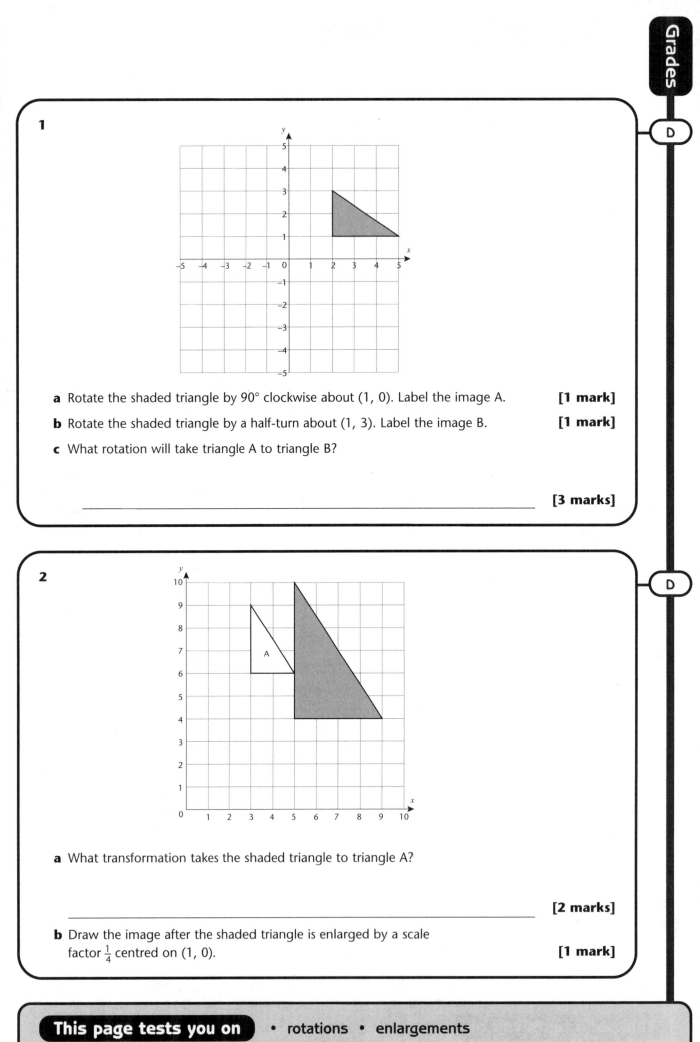

1

a Rotate the shaded triangle by 90° clockwise about (1, 0). Label the image A. **[1 mark]**

b Rotate the shaded triangle by a half-turn about (1, 3). Label the image B. **[1 mark]**

c What rotation will take triangle A to triangle B?

_____ **[3 marks]**

2

a What transformation takes the shaded triangle to triangle A?

_____ **[2 marks]**

b Draw the image after the shaded triangle is enlarged by a scale factor $\frac{1}{4}$ centred on (1, 0). **[1 mark]**

This page tests you on • rotations • enlargements

Constructions

D

1 Make an accurate drawing of this triangle.

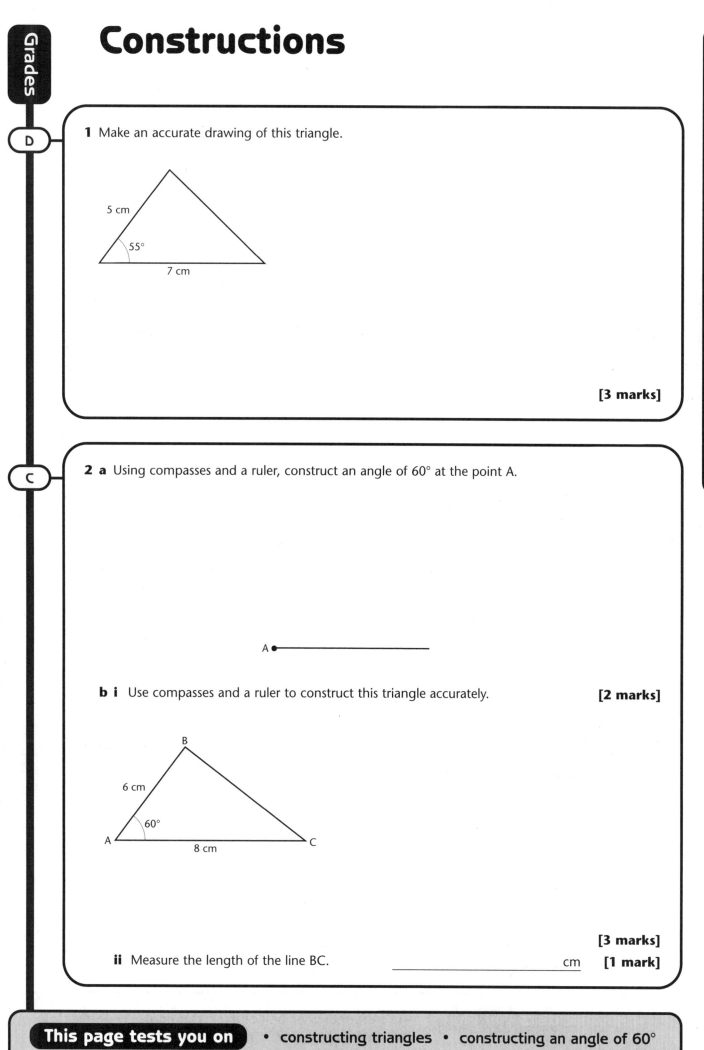

[3 marks]

C

2 a Using compasses and a ruler, construct an angle of 60° at the point A.

A •————————————

b i Use compasses and a ruler to construct this triangle accurately. [2 marks]

[3 marks]

ii Measure the length of the line BC. _____ cm [1 mark]

This page tests you on • constructing triangles • constructing an angle of 60°

1 Use compasses and a ruler to do these constructions.

   a Construct the perpendicular bisector of AB.

A ●

● B

[2 marks]

   b Construct the perpendicular at the point C to the line L.

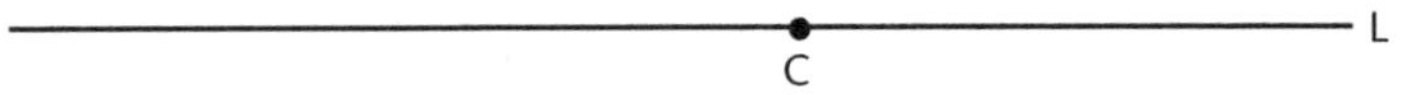

L

C

[2 marks]

C

2 Use compasses and ruler to do these constructions.

   a Construct the perpendicular bisector of the line L.

L

[2 marks]

   b Construct the bisector of angle ABC.

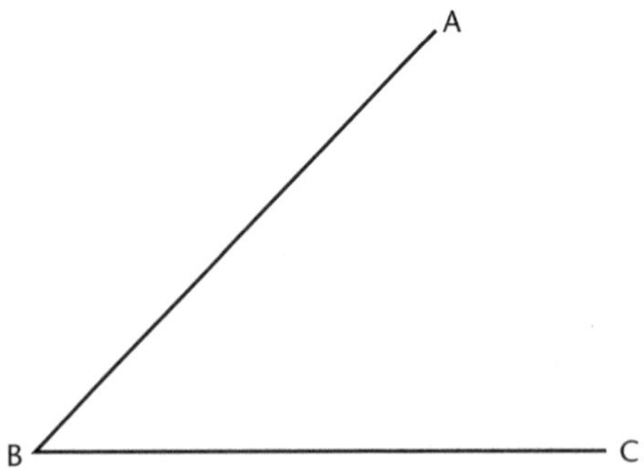

A

B   C

[2 marks]

C

This page tests you on   • the perpendicular bisector   • the angle bisector
                          • the perpendicular at a point on a line

Constructions and loci

C

1 Use compasses and a ruler to construct the perpendicular from the point C to the line L.

• C

———————————————————————————— L

[2 marks]

C

2 The diagram, which is drawn to scale, shows a flat, rectangular lawn of length 10 m and width 6 m, with a circular flower bed of radius 2 m.

Scale: 1 cm represents 1 m.

A tree is going to be planted in the garden.

It has to be at least 1 metre from the edge of the garden and at least 2 metres from the flower bed.

a Draw a circle to show the area around the flower bed where the tree *cannot* be planted.

[1 mark]

b Show the area of the garden in which the tree *can* be planted.

[1 mark]

This page tests you on
- the perpendicular from a point to a line
- loci • practical problems

Units

1 An old water butt is labelled: 'When full contains 50 gallons'.

Mary has a watering can that holds 9 litres.

a How many centilitres is 9 litres?

_____ cl   [1 mark]

b Approximately how many times can Mary fill the watering can from the water butt when it is full?

_____   [2 marks]

c A bottle of weedkiller says: 'Mix 200 g with 10 litres of water.'

How many grams of weedkiller will Mary have to mix with 9 litres of water?

_____ g   [2 marks]

F

2 a The safety instructions for Ahmed's trailer say:

'Load not to exceed 150 kg.'

Ahmed wants to carry 12 bags of sand, each of which weighs 30 lbs.

Can he carry them safely on the trailer?

_____   [1 mark]

b Ahmed has to drive 160 kilometres.

i How many metres is 160 kilometres?

_____ m   [1 mark]

ii Approximately how many miles is 160 kilometres?

_____ miles   [1 mark]

c Ahmed's car travels 30 miles to the gallon.

His tank contains 20 litres.

Will he have enough fuel in the tank to drive 160 kilometres?

_____   [2 marks]

F

This page tests you on   • systems of measurement • the metric system
• the imperial system • conversion factors

Surface area and volume of 3-D shapes

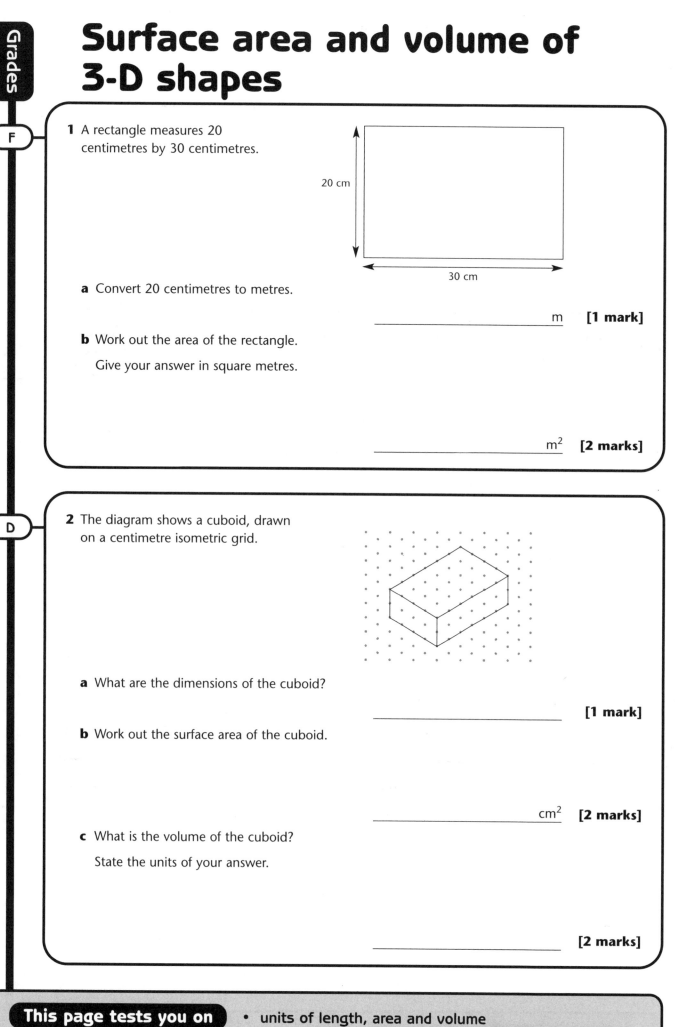

F

1 A rectangle measures 20 centimetres by 30 centimetres.

20 cm

30 cm

a Convert 20 centimetres to metres.

_____ m   **[1 mark]**

b Work out the area of the rectangle.

Give your answer in square metres.

_____ m²   **[2 marks]**

D

2 The diagram shows a cuboid, drawn on a centimetre isometric grid.

a What are the dimensions of the cuboid?

_____   **[1 mark]**

b Work out the surface area of the cuboid.

_____ cm²   **[2 marks]**

c What is the volume of the cuboid?

State the units of your answer.

_____   **[2 marks]**

This page tests you on
- units of length, area and volume
- surface area of a cuboid • volume of a cuboid

Density and prisms

1 A block of metal has a volume of 750 cm^3.

It has a mass of 5.1 kg.

Calculate the density of the metal.

State the units of your answer.

_____ **[2 marks]**

C

2 A triangular prism has dimensions as shown.

4 cm

12 cm

7 cm

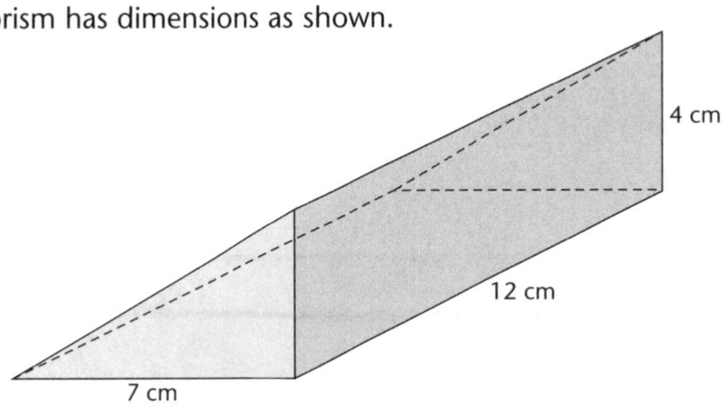

a Calculate the cross-sectional area of the prism.

_____ cm^2 **[1 mark]**

b Calculate the volume of the prism.

_____ cm^3 **[2 marks]**

D

3 A cylinder has a radius of 4 cm and a height of 10 cm.

What is the volume of the cylinder?

Give your answer in terms of π.

_____ cm^3 **[2 marks]**

C

This page tests you on • density • prisms and cylinders

Pythagoras' theorem

C

1 Calculate the length of the side marked x in this right-angled triangle.

Give your answer to 1 decimal place.

x

8 cm

10 cm

_____ cm **[2 marks]**

C

2 Calculate the length of the side marked x in this right-angled triangle.

Give your answer to 1 decimal place.

15 cm

x

11 cm

_____ cm **[2 marks]**

C

3 A flagpole 4 m tall is supported by a wire that is fixed at a point 2.1 m from the base of the pole.

How long is the wire?

(The length is marked x on the diagram.)

x

4 m

2.1 m

_____ m **[2 marks]**

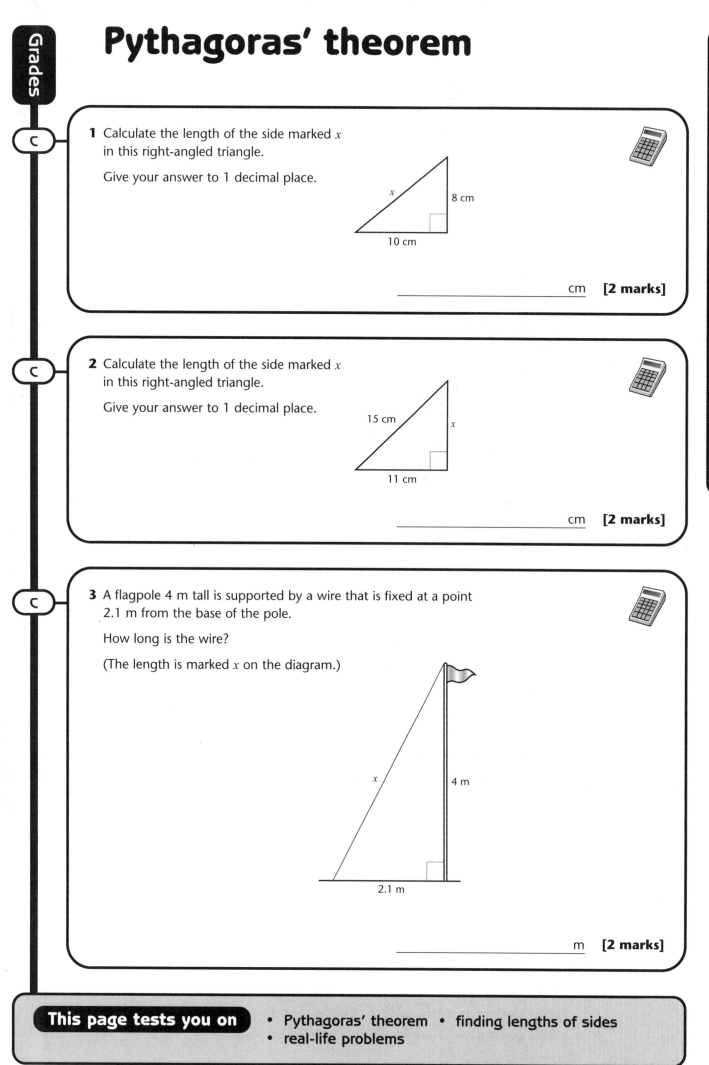

This page tests you on • Pythagoras' theorem • finding lengths of sides
• real-life problems

Shape, space and measures checklist

I can...

☐ find the perimeter of a 2-D shape

☐ find the area of a 2-D shape by counting squares

☐ draw lines of symmetry on basic 2-D shapes

☐ use the basic terminology associated with circles

☐ draw circles with a given radius

☐ recognise the net of a simple shape

☐ name basic 3-D solids

☐ recognise congruent shapes

☐ find the volume of a 3-D shape by counting squares

You are working at ⬭Grade G⬭ level.

☐ find the area of a rectangle, using the formula $A = lw$

☐ find the order of rotational symmetry for basic 2-D shapes

☐ measure and draw angles accurately

☐ use the facts that the angles on a straight line add up to 180° and the angles around a point add up to 360°

☐ draw and measure lines accurately

☐ draw the net of a simple 3-D shape

☐ read scales with a variety of divisions

☐ find the surface area of a 2-D shape by counting squares

You are working at ⬭Grade F⬭ level.

☐ find the area of a triangle using the formula $A = \frac{1}{2}bh$

☐ draw lines of symmetry on more complex 2-D shapes

☐ find the order of rotational symmetry for more complex 2-D shapes

☐ measure and draw bearings

☐ use the facts that the angles in a triangle add up to 180° and the angles in a quadrilateral add up to 360°

☐ find the exterior angle of a triangle and a quadrilateral

☐ recognise and find opposite angles

☐ draw simple shapes on an isometric grid

☐ tessellate a simple 2-D shape

☐ reflect a shape in the x- and y-axes

☐ convert from one metric unit to another

☐ convert from one imperial unit to another, given the conversion factor

☐ use the formula $V = lwh$ to find the volume of a cuboid

☐ find the surface area of a cuboid

You are working at ⬭Grade E⬭ level.

☐ find the area of a parallelogram, using the formula $A = bh$

☐ find the area of a trapezium, using the formula $\frac{1}{2}(a + b)h$

☐ find the area of a compound shape

- [ ] work out the formula for the perimeter, area or volume of simple shapes
- [ ] identify the planes of symmetry for 3-D shapes
- [ ] recognise and find alternate angles in parallel lines and a transversal
- [ ] recognise and find corresponding angles in parallel lines and a transversal
- [ ] recognise and find interior angles in parallel lines and a transversal
- [ ] use and recognise the properties of quadrilaterals
- [ ] find the exterior and interior angles of regular polygons
- [ ] understand the words 'sector' and 'segment' when used with circles
- [ ] calculate the circumference of a circle, giving the answer in terms of π if necessary
- [ ] calculate the area of a circle, giving the answer in terms of π if necessary
- [ ] recognise plan and elevation from isometric and other 3-D drawings
- [ ] translate a 2-D shape
- [ ] reflect a 2-D shape in lines of the form $y = a$, $x = b$
- [ ] rotate a 2-D shape about the origin
- [ ] enlarge a 2-D shape by a whole-number scale factor about the origin
- [ ] construct diagrams accurately, using compasses, a protractor and a straight edge
- [ ] use the appropriate conversion factors to change imperial units to metric units and vice versa

You are working at (Grade D) level.

- [ ] work out the formula for the perimeter, area or volume of complex shapes
- [ ] work out whether an expression or formula represents a length, an area or a volume
- [ ] relate the exterior and interior angles in regular polygons to the number of sides
- [ ] find the area and perimeter of semicircles
- [ ] translate a 2-D shape, using a vector
- [ ] reflect a 2-D shape in the lines $y = x$, $y = -x$
- [ ] rotate a 2-D shape about any point
- [ ] enlarge a 2-D shape by a fractional scale factor
- [ ] enlarge a 2-D shape about any centre
- [ ] construct perpendicular and angle bisectors
- [ ] construct an angle of 60°
- [ ] construct the perpendicular to a line from a point on the line and a point to a line
- [ ] draw simple loci
- [ ] work out the surface area and volume of a prism
- [ ] work out the volume of a cylinder, using the formula $V = \pi r^2 h$
- [ ] find the density of a 3-D shape
- [ ] find the hypotenuse of a right-angled triangle, using Pythagoras' theorem
- [ ] find the short side of a right-angled triangle, using Pythagoras' theorem
- [ ] use Pythagoras' theorem to solve real-life problems

You are working at (Grade C) level.

Basic algebra

1 The MacDonald family are Dad, Mum, Alfie and Bernice.

Alfie is x years old.

  a Bernice is two years younger than Alfie.

    Write down an expression for Bernice's age, in terms of x.

    _____ **[1 mark]**

  b Dad is twice as old as Alfie.

    Write down an expression for Dad's age, in terms of x.

    _____ **[1 mark]**

  c Mum is twice Bernice's age.

    Write down an expression for Mum's age, in terms of x.

    _____ **[1 mark]**

  d Write down and simplify an expression for the total age of the family, in terms of x.

    _____ **[2 marks]**

E

2 a Draw lines to show which algebraic expressions are equivalent.

   One line has been drawn for you.

	$3y$
$3y \times y$	$4y$
$3y + y$	$3y + 3$
$2y + y$	$5y + 2$
$3(y + 1)$	y^2
	$3y^2$

[3 marks]

  b Simplify each of these expressions.

    i $q + 4q - 2q$

    _____ **[1 mark]**

    ii $3p \times 5q$

    _____ **[1 mark]**

    iii $4x + 3 + 5x - 7$

    _____ **[1 mark]**

E

This page tests you on
• the language of algebra • simplifying expressions
• collecting like terms • multiplying expressions

Expanding and factorising

D

1 a Expand $5(x - 3)$.

_____ **[1 mark]**

b Expand and simplify $2(x + 1) + 2(3x + 2)$.

_____ **[2 marks]**

c A rectangle has length $2x + 3$ and width $x + 3$.

Write down and simplify an expression for the perimeter, in terms of x.

$x + 3$

$2x + 3$

_____ **[2 marks]**

D

2 a Multiply out and simplify $3(x - 4) + 2(4x + 1)$.

_____ **[2 marks]**

b Factorise each expression.

i $4x + 6$ **ii** $5x^2 + 2x$

_____   _____ **[1 mark each]**

C

3 a Expand and simplify each expression.

i $2x(3x - 4y) - x(x + 3y)$

_____ **[1 mark]**

ii $6(2x - 3y) - 2(x - 3y)$

_____ **[1 mark]**

b Complete these factorisations.

i $3xy^2 + 6x^2y = 3xy(\underline{} + \underline{})$ **ii** $4ab^2 - 8a^2b + 2a^2b^2$
$= 2ab(\underline{} - \underline{} + \underline{})$ **[1 mark each]**

c Factorise fully $3p^2q^2 + 6pq$.

_____ **[1 mark]**

This page tests you on   • expanding brackets • expand and simplify
• factorising

Quadratic expansion and substitution

1 a Expand $3(x + 2)$.

_____ **[1 mark]**

b Expand $x(x + 2)$.

_____ **[1 mark]**

c Expand and simplify $(x - 3)(x + 2)$.

_____ **[2 marks]**

d A rectangle has length $x + 2$ and width $x + 1$.

Write down an expression for the area, in terms of x, and simplify it.

$x + 1$

$x + 2$

_____ **[2 marks]**

C

2 a Multiply out and simplify $(x - 4)(x + 1)$.

_____ **[2 marks]**

b Multiply out and simplify $(x + 4)^2$.

_____ **[2 marks]**

C

3 a Work out the value of $3p + 2q$ when $p = -2$ and $q = 5$.

_____ **[1 mark]**

b Find the value of $a^2 + b^2$ when $a = 4$ and $b = 6$.

_____ **[1 mark]**

c An aeroplane has f first-class seats and e economy seats.

For a flight, each first-class seat costs £200 and each economy seat costs £50.

i If all seats are taken, write down an expression in terms of f and e for the total cost of all the seats in the aeroplane.

_____ **[1 mark]**

ii If $f = 20$ and $e = 120$, work out the actual cost of all the seats.

_____ **[1 mark]**

E

This page tests you on • quadratic expansion • squaring brackets • substitution

Linear equations

1 Solve these equations.

a $\frac{x}{3} + 5 = 4$

$x =$ _____ **[1 mark]**

b $3x + 4 = 1$

$x =$ _____ **[2 marks]**

c $\frac{7x - 2}{3} = 4$

$x =$ _____ **[2 marks]**

2 ABCD is a rectangle.

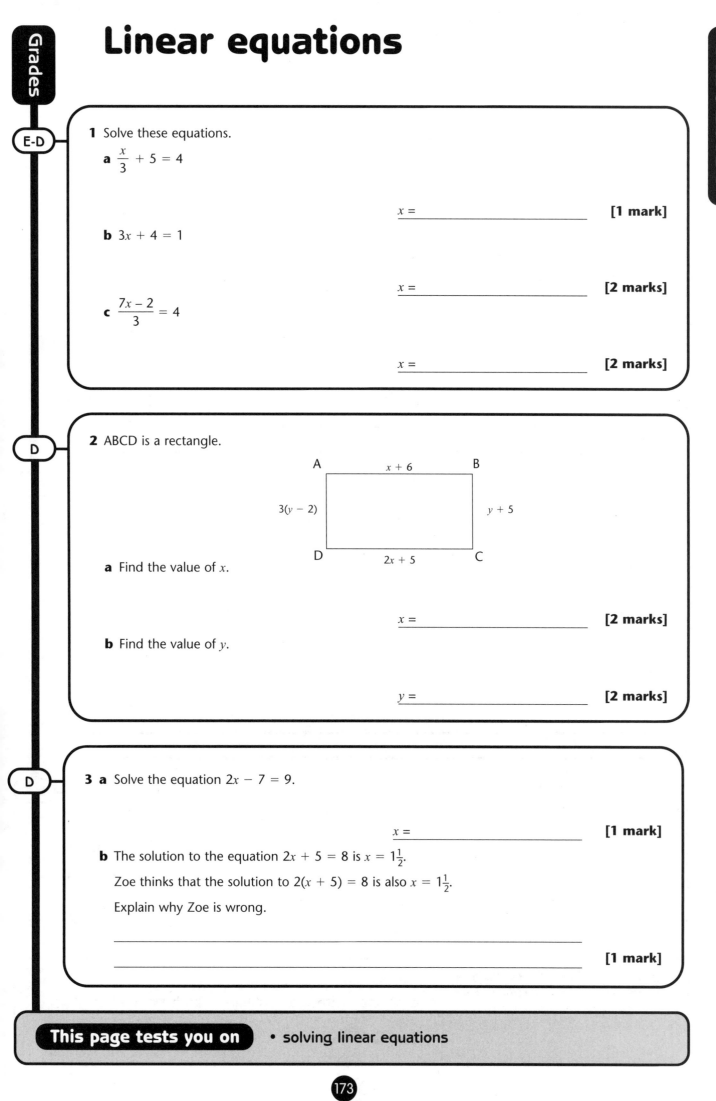

A — $x + 6$ — B
$3(y - 2)$ — — $y + 5$
D — $2x + 5$ — C

a Find the value of x.

$x =$ _____ **[2 marks]**

b Find the value of y.

$y =$ _____ **[2 marks]**

3 a Solve the equation $2x - 7 = 9$.

$x =$ _____ **[1 mark]**

b The solution to the equation $2x + 5 = 8$ is $x = 1\frac{1}{2}$.

Zoe thinks that the solution to $2(x + 5) = 8$ is also $x = 1\frac{1}{2}$.

Explain why Zoe is wrong.

_____

_____ **[1 mark]**

This page tests you on • solving linear equations

1 a I think of a number, multiply it by 3 and add 5. The answer is 26.

i Set up an equation to describe this.

_____ **[1 mark]**

ii Solve your equation to find the number.

_____ **[1 mark]**

b Solve these equations.

i $4(3y - 2) = 16$

$y =$ _____ **[2 marks]**

ii $5x - 2 = x + 10$

$x =$ _____ **[2 marks]**

E-D

2 Solve these equations.

a $5x - 2 = 3x + 1$

$x =$ _____ **[2 marks]**

b $3(x + 4) = x - 5$

$x =$ _____ **[2 marks]**

c $5(x - 2) = 2(x + 4)$

$x =$ _____ **[2 marks]**

D-C

3 Solve these equations.

a $4(x + 3) = x + 3$

$x =$ _____ **[2 marks]**

b $5(2x - 1) = 2(x - 3)$

$x =$ _____ **[2 marks]**

D-C

This page tests you on
- solving equations with brackets
- equations with the variable on both sides of the equals sign
- equations with brackets and the variable on both sides

E

1 a In the table, a, b, c and d each represent a different number.

The total of each row is shown at the side of the table.

a	a	a	a	16
a	a	b	b	20
a	a	b	c	21
a	b	c	d	25

Find the values of a, b, c and d.

$a =$ _____

$b =$ _____

$c =$ _____

$d =$ _____   **[2 marks]**

b i Write down an expression for the cost of x ice-lollies at 50p each and two choc-ices at 70p each.

_____   **[1 mark]**

ii The total cost of the x lollies and the two ice-lollies is £3.40.

Work out the value of x.

$x =$ _____   **[2 marks]**

C

2 The triangle has sides, given in centimetres, of x, $3x - 1$ and $2x + 5$.

The perimeter of the triangle is 25 cm.

Find the value of x.

$x =$ _____   **[3 marks]**

This page tests you on   • setting up equations

Trial and improvement

1 Use trial and improvement to solve the equation $x^3 + 4x = 203$.

The first two entries of the table are filled in.
Complete the table to find the solution.

Give your answer to 1 decimal place.

Guess	$x^3 + 4x$	Comment
5	145	Too low
6	240	Too high

$x =$ _____ **[3 marks]**

2 Darlene is using trial and improvement to find a solution to

$$2x + \frac{2}{x} = 8$$

The table shows her first trial.
Complete the table to find the solution.

Give your answer to 1 decimal place

Guess	$2x + \dfrac{2}{x}$	Comment
3	6.66	Too low

$x =$ _____ **[4 marks]**

This page tests you on • trial and improvement

Formulae

E

1 A widget weighs x grams.

A whotsit weighs 6 grams more than a widget.

a Write down an expression, in terms of x, for the weight of a whotsit.

_____ grams **[1 mark]**

b Write down an expression, in terms of x, for the total weight of three widgets and one whotsit.

_____ grams **[1 mark]**

c The total weight of three widgets and one whotsit is 27 grams.

Work out the weight of a widget.

_____ grams **[2 marks]**

C

2 a Explain why $5n - n \equiv 4n$ is an identity.

_____

_____ **[1 mark]**

b Explain why the equation $5(x + 1) - (x + 1) = 4(x + 1)$ does not have a solution.

_____

_____ **[1 mark]**

C

3 Rearrange each of these formulae to make x the subject.

a $C = \pi x$

$x =$ _____ **[1 mark]**

b $6y = 3x - 9$

Simplify your answer as much as possible.

$x =$ _____ **[2 marks]**

This page tests you on • formulae, identities, expressions and equations
• rearranging formulae

Inequalities

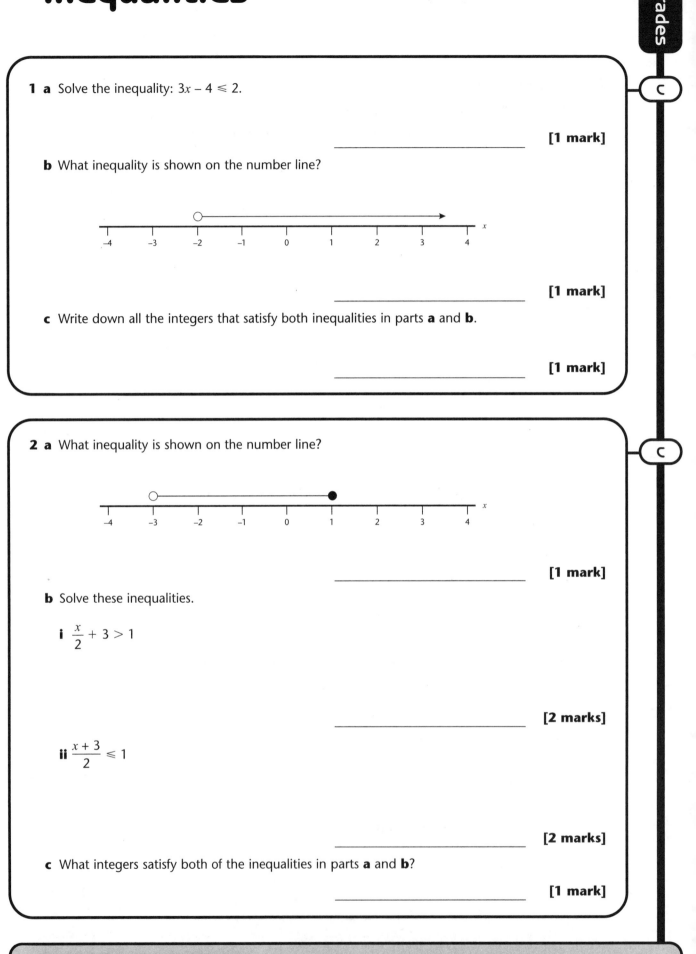

1 a Solve the inequality: $3x - 4 \leqslant 2$.

_____ **[1 mark]**

b What inequality is shown on the number line?

_____ **[1 mark]**

c Write down all the integers that satisfy both inequalities in parts **a** and **b**.

_____ **[1 mark]**

2 a What inequality is shown on the number line?

_____ **[1 mark]**

b Solve these inequalities.

i $\dfrac{x}{2} + 3 > 1$

_____ **[2 marks]**

ii $\dfrac{x + 3}{2} \leqslant 1$

_____ **[2 marks]**

c What integers satisfy both of the inequalities in parts **a** and **b**?

_____ **[1 mark]**

This page tests you on • inequalities • solving inequalities
• inequalities on number lines

Graphs

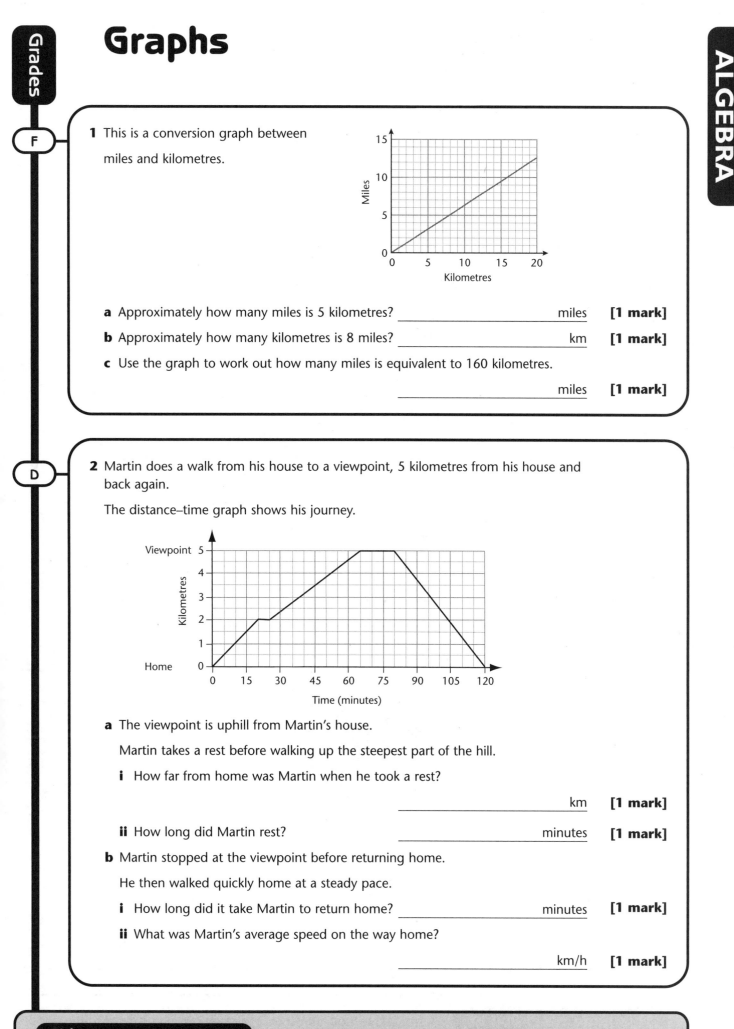

F

1 This is a conversion graph between miles and kilometres.

   a Approximately how many miles is 5 kilometres? _____ miles   **[1 mark]**

   b Approximately how many kilometres is 8 miles? _____ km   **[1 mark]**

   c Use the graph to work out how many miles is equivalent to 160 kilometres.

     _____ miles   **[1 mark]**

D

2 Martin does a walk from his house to a viewpoint, 5 kilometres from his house and back again.

The distance–time graph shows his journey.

   a The viewpoint is uphill from Martin's house.

     Martin takes a rest before walking up the steepest part of the hill.

     i How far from home was Martin when he took a rest?

       _____ km   **[1 mark]**

     ii How long did Martin rest? _____ minutes   **[1 mark]**

   b Martin stopped at the viewpoint before returning home.

     He then walked quickly home at a steady pace.

     i How long did it take Martin to return home? _____ minutes   **[1 mark]**

     ii What was Martin's average speed on the way home?

       _____ km/h   **[1 mark]**

This page tests you on   • conversion graphs • travel graphs

Linear graphs

1 The table shows some values of the function $y = 3x + 1$ for values of x from –1 to 3.

a Complete the table of values.

x	–1	0	1	2	3
y	–2				10

[1 mark]

b Use the table to draw the graph of $y = 3x + 1$.

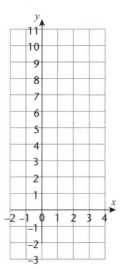

[2 marks]

c What is the x-value when $y = 8$? _____ **[1 mark]**

2 Draw the graph of $y = 2x - 1$ for $-3 \leqslant x \leqslant 3$.

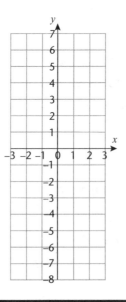

[2 marks]

This page tests you on   • negative coordinates • drawing graphs from tables
  • drawing linear graphs

C

1 Here are the equations of six lines.

A: $y = 3x + 6$ B: $y = 2x - 1$ C: $y = \dfrac{x}{2} - 1$

D: $y = 3x + 1$ E: $y = \dfrac{x}{3} + 1$ F: $y = 4x + 2$

a Which line is parallel to line A?

_____ **[1 mark]**

b Which line crosses the y-axis at the same point as line B?

_____ **[1 mark]**

c Which other two lines intersect on the y-axis?

_____ **[1 mark]**

d Write down the gradient of each of these lines.

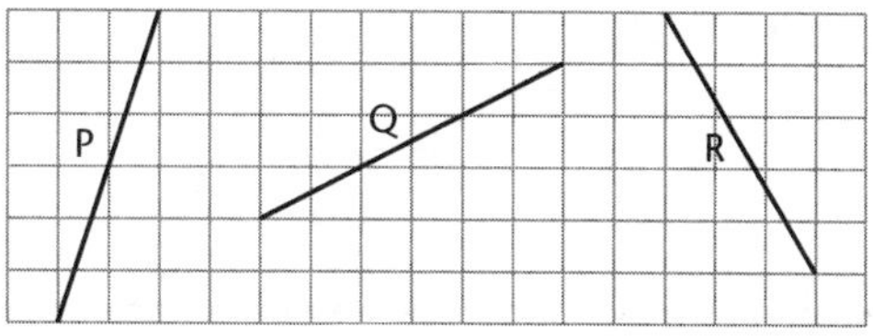

Line P _____ Line Q _____ Line R _____ **[3 marks]**

C

2 Use the gradient-intercept method to draw the graph of $y = 3x - 2$ for $-3 \leqslant x \leqslant 3$.

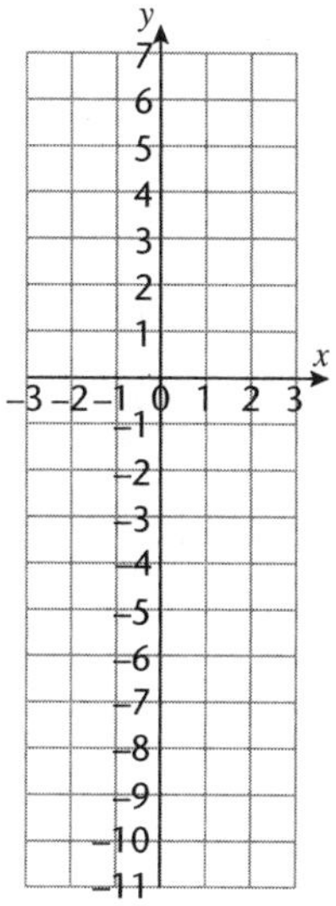

[2 marks]

This page tests you on   • gradients • the gradient–intercept method
• drawing a line with a given gradient

Quadratic graphs

1 a Complete the table of values for the graph of $y = x^2 + 3$.

x	−3	−2	−1	0	1	2	3
y	12	7					12

[1 mark]

b Draw the graph of $y = x^2 + 3$ for values of x from −3 to 3.

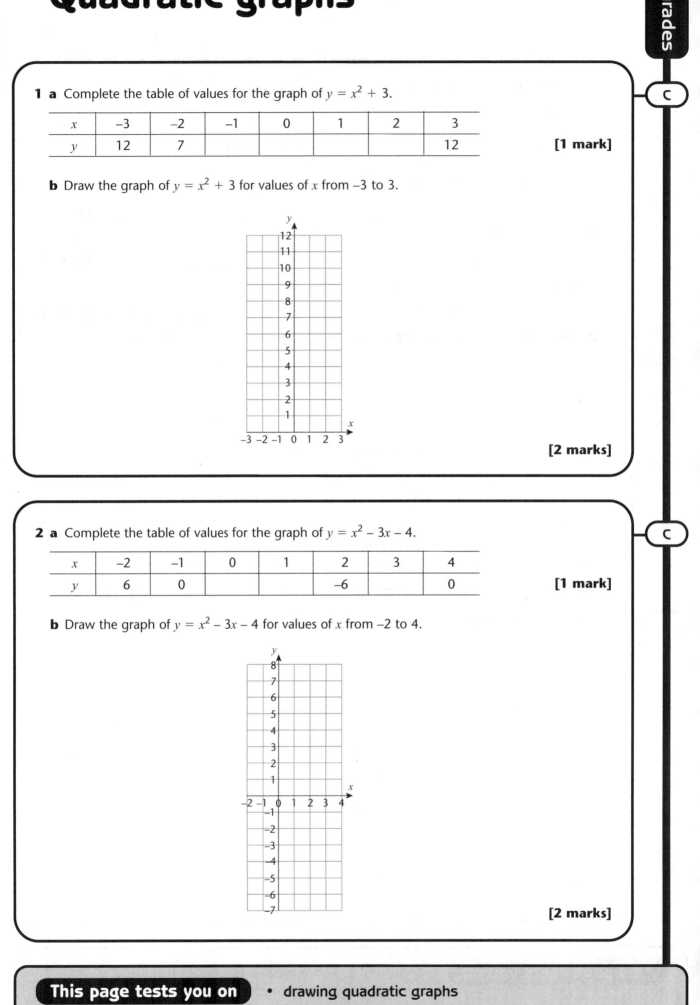

[2 marks]

2 a Complete the table of values for the graph of $y = x^2 - 3x - 4$.

x	−2	−1	0	1	2	3	4
y	6	0			−6		0

[1 mark]

b Draw the graph of $y = x^2 - 3x - 4$ for values of x from −2 to 4.

[2 marks]

This page tests you on • drawing quadratic graphs

C

1 a Complete the table of values for the graph of $y = x^2 - 2x + 1$.

x	-2	-1	0	1	2	3	4
y	9	4	1				

[1 mark]

b Draw the graph of $y = x^2 - 2x + 1$ for values of x from -2 to 4.

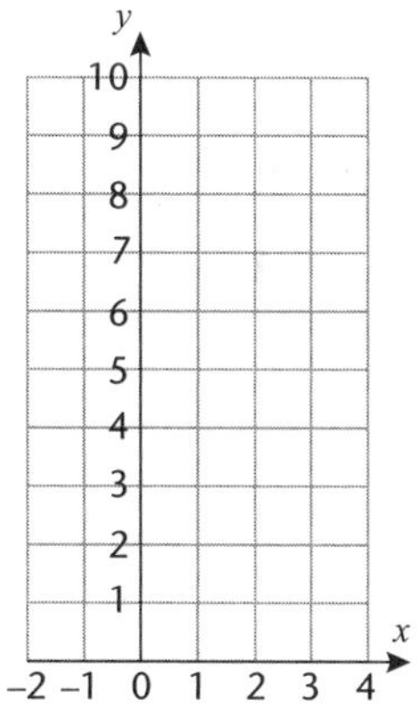

[2 marks]

c Use the graph to find the x-values when $y = 6$. _____ [1 mark]

d Use the graph to solve the equation $x^2 - 2x + 1 = 0$.

_____ [1 mark]

C

2 a Complete the table of values for the graph of $y = x^2 + 2x - 1$.

x	-4	-3	-2	-1	0	1	2
y		2	-1				7

[1 mark]

b Draw the graph of $y = x^2 + 2x - 1$ for values of x from -4 to 2.

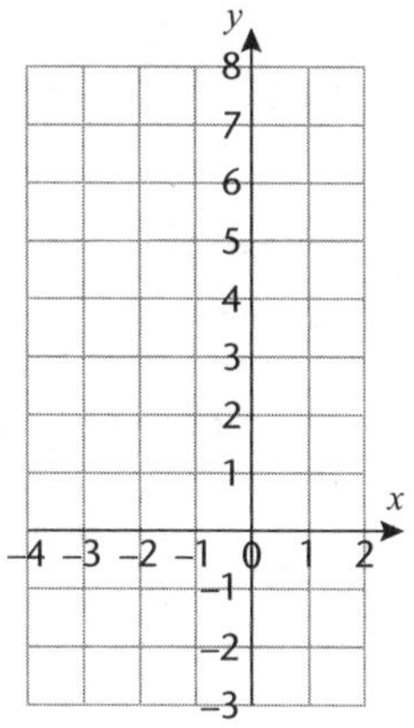

[2 marks]

c Use the graph to find the x-values when $y = 1.5$. _____ [1 mark]

d Use the graph to solve the equation $x^2 + 2x - 1 = 0$.

_____ [1 mark]

This page tests you on
• reading values from quadratic graphs
• using graphs to solve quadratic equations

Pattern

1 a Here are three lines of a series of number calculations.

1	=	1	=	1^2
1 + 3	=	4	=	2^2
1 + 3 + 5	=	9	=	3^2
___	=	___	=	___
___	=	___	=	___

Complete the next two lines of the pattern. **[2 marks]**

b 1, 3, 5, 7, 9, 11, ... are the **odd numbers**.

What is the 50th odd number?

_____ **[1 mark]**

c 1, 4, 9, ... are the **square numbers**.

What is the 15th square number?

_____ **[1 mark]**

2 Squares are used to make patterns.

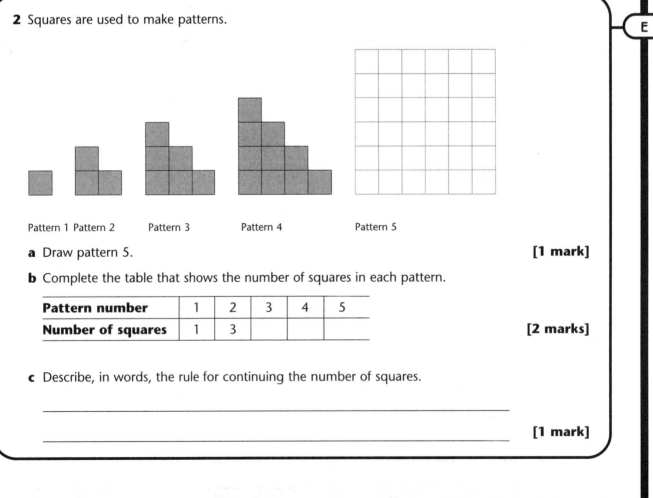

Pattern 1 Pattern 2   Pattern 3   Pattern 4   Pattern 5

a Draw pattern 5. **[1 mark]**

b Complete the table that shows the number of squares in each pattern.

Pattern number	1	2	3	4	5
Number of squares	1	3			

[2 marks]

c Describe, in words, the rule for continuing the number of squares.

_____

_____ **[1 mark]**

This page tests you on • patterns in number • number sequences

The *n*th term

1 The *n*th term of a sequence is $4n + 1$.

   a Write down the first three terms of the sequence.

      _____ **[1 mark]**

   b Which term of the sequence is equal to 29?

      _____ **[1 mark]**

   c Explain why 84 is not a term in this sequence.

      _____

      _____ **[1 mark]**

   d What is the *n*th term of the sequence 3, 10, 17, 24, 31, __?

      _____ **[2 marks]**

2 Matches are used to make patterns with hexagons.

Pattern 1   Pattern 2   Pattern 3   Pattern 4

   a Complete the table that shows the number of matches used to make each pattern.

Pattern number	1	2	3	4	5
Number of squares	6	11			

[2 marks]

   b How many matches will be needed to make the 20th pattern?

      _____ **[1 mark]**

   c How many matches will be needed to make the *n*th pattern?

      _____ **[2 marks]**

This page tests you on   • the *n*th term of a sequence   • finding the *n*th term

Sequences

D

1 R is an odd number, Q is an even number, P is a prime number.

State whether these expressions are *always even*, *always odd* or *could be either*.

	Always even	Always odd	Could be either	
a $R + Q$	☐	☐	☐	[1 mark]
b RQ	☐	☐	☐	[1 mark]
c $P + Q$	☐	☐	☐	[1 mark]
d R^2	☐	☐	☐	[1 mark]
e $R + PQ$	☐	☐	☐	[1 mark]

C

2 a n is a positive integer.

Explain why $2n$ is always an even number.

_____

_____ [1 mark]

b Zoe says that when you square an even number you always get a multiple of 4.
Show that Zoe is correct.

_____

_____ [2 marks]

E

3 Triangles are used to make patterns.

Pattern 1   Pattern 2   Pattern 3

a Complete the table that shows the number of triangles used to make each pattern.

Pattern number	1	2	3	4	5
Number of triangles	12				

[2 marks]

b How many triangles will be needed to make the nth pattern?

_____ [2 marks]

This page tests you on
- special sequences
- finding the nth term from given patterns

Algebra checklist

I can...

- [ ] use a formula expressed in words
- [ ] substitute numbers into expressions
- [ ] use letters to write a simple algebraic expression
- [ ] solve linear equations that require only one inverse operation to solve
- [ ] read values from a conversion graph
- [ ] plot coordinates in all four quadrants
- [ ] give the next value in a linear sequence
- [ ] describe how a linear sequence is building up

You are working at (Grade F) level.

- [ ] simplify an expression by collecting like terms
- [ ] simplify expressions by multiplying terms
- [ ] solve linear equations that require more than one inverse operation to solve
- [ ] read distances and times from a travel graph
- [ ] draw a linear graph from a table of values
- [ ] find any number term in a linear sequence
- [ ] recognise patterns in number calculations

You are working at (Grade E) level.

- [ ] use letters to write more complicated algebraic expressions
- [ ] expand expressions with brackets
- [ ] factorise simple expressions
- [ ] solve linear equations where the variable appears on both sides of the equals sign
- [ ] solve linear equations that require the expansion of a bracket
- [ ] set up and solve simple equations from real-life situations
- [ ] find the average speed from a travel graph
- [ ] draw a linear graph without a table of values
- [ ] substitute numbers into an nth term rule
- [ ] understand how odd and even numbers interact in addition, subtraction and multiplication problems

You are working at (Grade D) level.

- [ ] expand and simplify expressions involving brackets
- [ ] factorise expressions involving letters and numbers
- [ ] expand pairs of linear brackets to give a quadratic expression
- [ ] solve linear equations that have the variable on both sides and include brackets
- [ ] solve simple linear inequalities
- [ ] show inequalities on a number line
- [ ] solve equations, using trial and improvement
- [ ] rearrange simple formulae
- [ ] use a table of values to draw a simple quadratic graph
- [ ] use a table of values to draw a more complex quadratic graph
- [ ] solve a quadratic equation from a graph
- [ ] give the nth term of a linear sequence
- [ ] give the nth term of a sequence of powers of 2 or 10.

You are working at (Grade C) level.

Formulae sheet

Area of trapezium $= \frac{1}{2}(a + b)h$

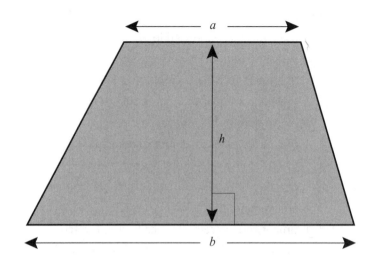

Volume of prism = acrea of cross-section × length

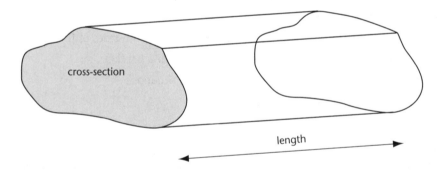

Notes

Notes

Notes

Notes

Answers

Handling data

Page 100 Statistical representation

1 a 13; 6; 2

   b

   c 50

   d Because 21 cars had 1 passenger (21); 13 cars had 2 passengers (26); 8 cars had 3 passengers (24); 6 cars had 4 passengers (24); 2 cars had 5 passengers (10) and 21 + 26 + 24 + 24 + 10 (= 105)

                 (1 mark for a partial explanation)

2 a 95; 70

   b 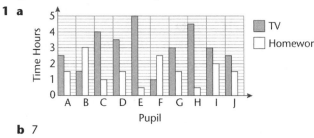 and

   c 305

Page 101 Statistical representation

1 a

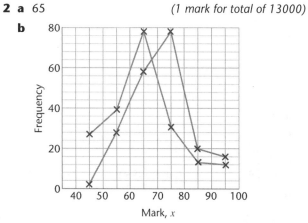

   b 7

   c No, student B and student F spent more time doing homework than watching TV

2 a 16 °C

   b 10am–11am

   c 11am

   d 25 °C

   e No, anything could happen in four hours, it is too far from the end of the graph

Page 102 Averages

1 a 50

   b 1

   c 2

   d 54%

2 a i 64

    ii 25

   b i 924 kg

    ii 98 kg   *(1 mark for 14 3 87 – 924 (= 294))*

Page 103 Averages

1 a 24

   b 2

   c 2

   d 1.75   *(1 mark for total = 42)*

   e No; they will need 420 spoons

2 a i 46

    ii 98

    iii 106

   b i Only 2 people and the lowest value

    ii Affected by one extreme value

   c No; mode does not tell you anything about the total score

   d i Yes; total is now 1200 rather than 1060

    ii Cannot tell as 5 decreased their score

Page 104 Arranging data

1 a 2

   b 2

   c 1.8   *(1 mark for total of 180)*

2 a 65   *(1 mark for total of 13000)*

   b

   c Girls did better; polygons are about the same shape and girls are about 10 marks better

1

Page 105 Arranging data

1 a

0	7 8 9 9
1	1 2 3 3 3 3 5 6 8 9
2	1 2 2 3 4 4

  b **i** 130

    ii 140

    iii 170

2 a 30

  b 9.6   *(1 mark for 288)*

Page 106 Probability

1 a i Impossible

    ii Very unlikely

    iii Even

    iv Very likely/certain

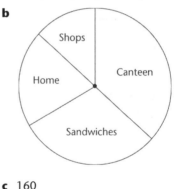

2 a i $\frac{12}{20} = \frac{3}{5}$

    ii 0

    iii $\frac{8}{20} = \frac{2}{5}$

  b 4

Page 107 Probability

1 a $\frac{47}{50}$

  b 30

2 a Mrs Rogers; she has more tickets

  b $\frac{5}{900} = \frac{1}{180}$

  c $\frac{4}{900} = \frac{1}{225}$

  d $\frac{19}{900}$

3 a 0.15, 0.06, 0.18, 0.18, 0.29, 0.14

                    (1 mark for 4 or 5 correct)

  b 5; as this had a much higher relative frequency

Page 108 Probability

1 a Sausage, hash browns and beans; sausage, hash browns and toast; bacon, eggs and beans; bacon, eggs and toast; bacon, hash browns and beans; bacon, hash browns and toast

  b $\frac{2}{8} = \frac{1}{4}$

2 a 12

  b **i** $\frac{3}{12} = \frac{1}{4}$

    ii $\frac{2}{12} = \frac{1}{6}$

  c **i** Top row: 2, 4, 6, 8, 10, 12; bottom row: 0, 1, 2, 3, 4, 5

    ii $\frac{3}{12} = \frac{1}{4}$

Page 109 Probability

1 a As 2 + 3 = 5; the probability of red is $\frac{2}{5}$

  b 12

  c 80

2 a 22   **b** PE

  c Maths $\frac{5}{12} \approx 42\%$; Science $\frac{7}{18} \approx 39\%$

  d $\frac{12}{40} = \frac{3}{10}$

Page 110 Pie charts

1 a 132°; 108°; 72°; 48°   *(1 mark for any frequency 3 6)*

  b

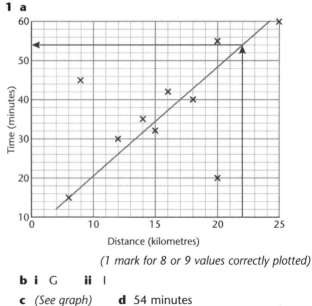

                        (1 mark for sectors)

                        (1 mark for labels)

  c 160

Page 111 Scatter diagrams

1 a

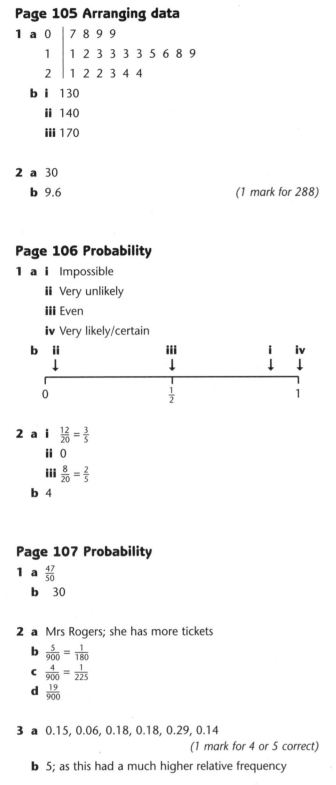

    (1 mark for 8 or 9 values correctly plotted)

  b **i** G   **ii** I

  c *(See graph)*   **d** 54 minutes

Page 112 Surveys

1 a

	Boys	Girls
0 < time (hours) ≤ 4		
4 < time (hours) ≤ 6		
6 < time (hours) ≤ 8		
8 < time (hours) ≤ 10		
More than 10 hours		

(1 mark for times or for boys/girls)

b Not really; the difference is not that large and the sample of girls was too small

2 a 94.80p

b 2006; price index shot up

c Faster; petrol has increased by 58%

Number

Page 114 Basic number

1 a i 72

ii 280

iii 80

b i $4 \times 7 = 28$

ii $28 \div 7 = 4$ or $28 \div 4 = 7$

2 a 4

b i $(2 + 4^2) \div 8 = 2.25$

ii $(2 + 4)^2 \div 8 = 4.5$

Page 115 Basic number

1 a 3659

b 6864

c 1770

2 a i Ninety-six thousand, nine hundred and twenty-four

ii 96 900

iii 97 000

b i 9 thousand

ii 68 950

iii 69 049

iv 23 000

3 a 898

b 3178

c 1225

4 a 1139

b 2317

Page 116 Basic number

1 a 304

b 26

c 378

d 48

2 a i 272p or £2.72

ii 228p or £2.28

b 720

c 28

3 a $20 - 12.85 = £7.15$

b $14.55 \div 3 = £4.85$

c $9 \times 12 = 108$

Page 117 Fractions

1 a $\frac{9}{12}$ and $\frac{60}{80}$

b Any 18 squares shaded

c i $\frac{14}{20}$ or $\frac{7}{10}$

ii $\frac{6}{20}$ or $\frac{3}{10}$

2 a i Any 2 squares shaded

ii Any 3 squares shaded

b $\frac{5}{9}$

3 a The fraction is $\frac{2}{5}$

b $\frac{4}{10}$ or $\frac{6}{15}$ or $\frac{20}{50}$ or $\frac{8}{20}$ etc.

Page 118 Fractions

1 a i Any 2 squares shaded

ii Any 4 squares shaded

iii Any 6 squares shaded

iv Any 10 squares shaded   *(1 mark for 3 correct)*

b i 12   **ii** 32   **iii** 6   **iv** 2

c i $\frac{4}{7}$ **ii** $\frac{3}{5}$ **iii** $\frac{2}{5}$ **iv** $\frac{3}{4}$

d $\frac{13}{20}; \frac{7}{10}; \frac{4}{5}$

2 a i $1\frac{4}{5}$ **ii** $2\frac{3}{7}$ **iii** $2\frac{5}{8}$ **iv** $7\frac{3}{4}$

b i $\frac{17}{11}$ **ii** $\frac{11}{8}$

iii $\frac{7}{3}$ **iv** $\frac{23}{5}$

Page 119 Fractions

1 a i 6

   ii 10

   iii 2

   iv 9

 b i $\frac{7}{20}$

   ii $\frac{13}{16}$

   iii $\frac{1}{2}$

   iv $\frac{5}{6}$

 c i $\frac{11}{20}$

   ii $\frac{5}{16}$

   iii $\frac{1}{2}$

   iv $\frac{1}{6}$

2 a i $\frac{3}{5}$ **ii** 98

 b 210   *(1 mark for 90 × 2 + 30)*

 c 165   *(1 mark for 55)*

Page 120 Fractions

1 a i $\frac{4}{11}$

   ii $\frac{1}{6}$

   iii $\frac{1}{8}$

   iv $\frac{3}{4}$

 b i $1\frac{1}{2}$

   ii $1\frac{1}{2}$

   iii 2

   iv 6

2 a $\frac{1}{6}$ *(1 mark for $\frac{250}{1500}$)*

 b $\frac{1}{3}$ *(1 mark for $\frac{40}{120}$)*

3 a £2700   *(1 mark for £900)*

 b 1.4 kg   *(1 mark for 200 g or 0.2 kg)*

Page 121 Rational numbers and reciprocals

1 a i 0.175

   ii 0.7$\dot{3}$

   iii 0.8$\dot{3}$

   iv 0.18

 b i 0.4444...   **ii** 0.5555...

 c i 0.2727....   **ii** 0.5454...

2 a i $\frac{1}{10}$ **ii** $1\frac{1}{3}$

 b i 0.1   **ii** 1.$\dot{3}$

 c i 0.8   **ii** 0.4   **iii** 0.2

 d Keep halving the reciprocals, i.e. 0.2, 0.1, 0.05,
    0.025   *(1 mark for mentioning halving)*

Page 122 Negative numbers

1 a Aberdeen

 b 9°

 c Aberdeen and Bristol

 d London

2 a

 b −2.2; −2.1; −2.0

 c i 2

   ii −9.5

Page 123 Negative numbers

1 a −1

 b +6; −6

 c i +8

   ii −4

 d i −7

   ii 11

2 −11, +2, −2, −1   (1 mark for 3 correct)

Page 124 More about number

1 a 6 and 12

b 6 and 10

c

12	7	8
5	9	13
10	11	6

2 a i {1, 3, 11, 33}

ii {1, 2, 3, 6, 9, 18}

b i 84 or 90

ii 85 or 90

c 60 seconds

Page 125 Primes and squares

1 a 11 and 13

b 2 is even and prime

c 9 and 16

d $5^2 = 25$; $10^2 = 100$

e 11 and 13

2 a Either odd or even

b Either odd or even

c Always odd

d Either odd or even

Page 126 Roots and powers

1 a $\sqrt{49} = 7$

b $\sqrt{64} = 8$

c $2^4 = 16 > \sqrt{144} = 12$

d i 13

ii 125

e i 4

ii 256

2 a 64, 4; 256, 6; 1024, 4

b 4; all odd powers of 4 end in 4

c $5^6 = 15625 > 6^5 = 7776$

Page 127 Powers of 10

1 a 4 tenths; 0.4 or $\frac{4}{10}$

b 10^4

c 10 000 000

d 100; $\frac{1}{100}$; 10^1; 10^0; 10^{-3} *(1 mark for 4 correct)*

e i 1

ii 5

2 a i 370

ii 250

b i 0.76

ii 0.0065

c i 12 000 000

ii 360 000

d i 3000

ii 500

Page 128 Prime factors

1 a 90

b $2 \times 5 \times 7$

c $2^4 \times 3$

d i 9, 3, 3, 10, 2, 5, 2, 5   *(1 mark for 4 correct)*

ii $2^2 \times 3^2 \times 5^2$

2 a i 5   **ii** $2 \times 3 \times 5^2$

b i 3   **ii** $2^3 \times 3^3$

Page 129 LCM and HCF

1 a $2^3 \times 3$

b $2^2 \times 3 \times 5$

c 120

d 12

e i $2^4 \times 3^2 \times 5^2$ **ii** $2^2 \times 3 \times 5$

2 a $p = 2$, $q = 3$

b $2^3 \times 3^2 \times 5$

c $a = 2$, $b = 7$

d $2^2 \times 7^2$

Page 130 Powers

1 a 4^8

   b 6^3

   c i 4   **ii** 3

   d 49

   e 100 000 000

   f 1 000 000

2 a x^7

   b x^4

   c $(ab)^n$

   d $(a \div b)^n$

Page 131 Number skills

1 a i He has forgotten the zero when multiplying by the 4

   ii 1776

   b i He has added instead of multiplying

   ii 1363   *(1 mark for a partial method)*

2 a 3456   *(1 mark for a partial method)*

   b £23.52   *(1 mark for a partial method)*

Page 132 Number skills

1 a i 29

   ii 58

   iii 290

   iv 580   *(1 mark for 3)*

   b
```
    1 5 0 8
  -   5 8 0
      9 2 8
  -   5 8 0
      3 4 8
  -   2 9 0
        5 8
  -     5 8
         0
```
   1508 ÷ 29 = 52

   c i 52 remainder 12

   ii 26

2 a 14 592   *(1 mark for a partial method)*

   b 38   *(1 mark for a partial method)*

Page 133 Number skills

1 a 476   *(1 mark for a partial method)*

   b 9   *(1 mark for 8 remainder 12)*

2 a i 40 or 10

   ii 0.007 or $\frac{1}{1000}$ or 0.001

   b i 23.5

   ii 23.48

   iii 23.479

Page 134 Number skills

1 a £3.84 + £4.30 + £1.10 = £9.24

   b i 7.8

   ii 6.24   *(1 mark for 1.04 and 5.2)*

   c i £6.24

   ii £300

2 a i 17.4

   ii 13.34   *(1 mark for 11.6 or 1.74)*

   b £13.34

Page 135 More fractions

1 a $1\frac{3}{20}$ *(1 mark for $\frac{15}{20} + \frac{8}{20}$)*

   b $1\frac{13}{15}$ *(1 mark for $\frac{11}{3} - \frac{9}{5}$)*

   c $\frac{11}{60}$ *(1 mark for $\frac{49}{60}$)*

2 a $3\frac{1}{2}$ *(1 mark for $\frac{5}{2} \times \frac{7}{5}$)*

   b $1\frac{3}{8}$ *(1 mark for $\frac{33}{10} \times \frac{5}{12}$)*

   c $\frac{39}{40}$ *(1 mark for $\frac{1}{2} \times \frac{13}{6} \times \frac{9}{10}$)*

Page 136 More number

1 a i −12

   ii −3

   iii −20

   b Any two negatives where the second is 2 less than the first, e.g. −5 − −7; −9 − −11

   c i 9 is a square number but −4 is not; as all square numbers are positive

   ii 1   *(1 mark for total = 7)*

2 a i 50

   ii 0.4

   b 200   *(1 mark for (50 × 40) ÷ (20 − 10))*

   c 50   *(1 mark for (50 − 30) ÷ 0.4)*

Page 137 Ratio

1 360° *(1 mark)*; 360° ÷ 18 (= 20°) *(1 mark)*; largest angle = 160°

2 a 4 : 3

b 1 : 0.4

c 375 ml   *(1 mark for 1000 ÷ 8 or 125 ml)*

3 35   *(1 mark for 15 ÷ 3 (= 5))*

Page 138 Speed and proportion

1 a 32 km per hour

    (1 mark for 72 ÷ 2.25, 1 mark for units)

b 36 km per hour   *(1 mark for 63 km)*

2 a 30   *(1 mark for 5.5)*

b 143 miles

3 Handy size 3.87 g/p compared to large size 3.64 g/p

    (1 mark for grams ÷ pence)

Page 139 Percentages

1 a

Decimal	Fraction	Percentage
0.35	$\frac{7}{20}$	35
0.8	$\frac{4}{5}$	80
0.9	$\frac{9}{10}$	90

    (2 marks for 4 or 5 right; 1 mark for 2 or 3 right)

b i $\frac{3}{10}$

  ii 70%

c $\frac{11}{20}$ (0.55); $\frac{14}{25}$ (0.56); 57% (0.57); 0.6

2 a 6000 × 0.88 × 0.9

b £6.80 + £3.40 + £1.70 *(1 mark)*; £11.90

Page 140 Percentages

1 a £52.50   *(1 mark for 0.15 × 350)*

b £297.50

2 a £822.50   *(1 mark for 1.175 × 700)*

b £220   *(1 mark for 0.88 × 250)*

3 18%   *(1 mark for 405 ÷ 2250)*

Shape, space and measures
Page 143 Perimeter and area

1 a 274 *(1 mark)* cm *(1 mark)*

b 4602 *(1 mark)* cm2 *(1 mark)*

2 a 32–36 *(1 mark)* km2 *(1 mark)*

b 18 cm²

Page 144 Perimeter and area

1 100 cm²   *(1 mark for 60 seen in calculation)*

2 a 27 cm²   *(1 mark for $\frac{1}{2}$ × 9 × 6)*

b 87 cm²   *(1 mark for 60 + area of triangle)*

Page 145 Perimeter and area

1 a 24 cm²

b 44 cm²

2 152 cm²   *(1 mark for $\frac{1}{2}$ × (22 + 16) × 8)*

Page 146 Dimensional analysis

1 Area; volume; length

2 Area; perimeter; volume

Page 147 Symmetry

1 a

b

c i 0

  ii 2

2 a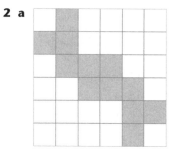

b 5

Page 148 Angles

1 a i 62°

ii 220°

b

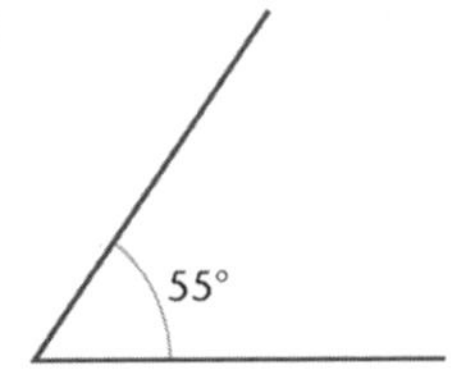

2 a 37°

b 157°

Page 149 Angles

1 a i Isosceles

ii $p = 72°$, $q = 36°$

b 133°

2 67° *(1 mark for 113°)*

Page 150 Polygons

1 a $p = 45°$, $q = 135°$, $r = 45°$

b A pentagon can be split into three triangles *(1 mark)*

$3 \times 180° = 540°$ *(1 mark)*

2 a 10 *(1 mark for 360° ÷ 36°)*

b Exterior angle = 20° *(1 mark)*

360° ÷ 20° = 1 *(1 mark)*

Page 151 Parallel lines and angles

1 a 45°; interior or allied angle

b 65°; corresponding angle

c 70°

2 $p = 38°$; opposite

$q = 38°$; alternate

$r = 38°$; corresponding

$s = 142°$; allied

Page 152 Quadrilaterals

1 a Parallelogram

b i Square

ii Rhombus

2 40° *(1 mark for ∠BCD = 100°; 1 mark for ∠DCE = 80°; 1 mark for ∠CDE = 60°)*

Page 153 Bearings

1 *Allow ±0.1 km and ±1°*

a 3.6 km at 034°

b 5.1 km at 169°

c 5 km at 233°

d 3.6 km at 304°

2 000° or 360°; 090°; 180°; 270°

Page 154 Circles

1 Top left: radius; bottom left: tangent; top right: diameter; bottom right: chord

2 a 113.1 cm^2

b 62.8 cm

3 50π cm^2

Page 155 Scales

1 a 28.4 °C

b 57 *(1 mark)* mph *(1 mark)*

c

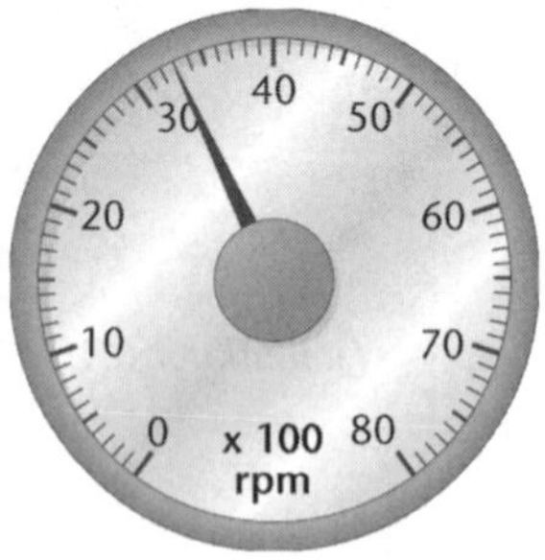

2 6 × height man (1.8 m) *(1 mark)* 10–12 m *(1 mark)*

Page 156 Scales and drawing

1 *Allow ±1 mm*

   a 6.7 cm

   b and **c**

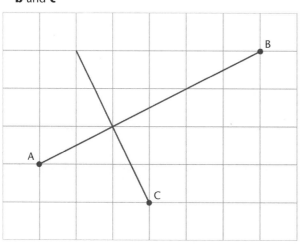

   d 5 cm *(1 mark)*; 25 km *(1 mark)*

2 a Triangular prism

   b 1.3 cm

   c Area rectangle 1.5 cm by 4 cm = 6 cm^2
     Area triangle $\frac{1}{2} \times 1.5 \times 1.3 = 0.975$ cm^2
     Total area = 19.5–20 cm^2

Page 157 3-D drawing

1 a 80 cm^3

   b

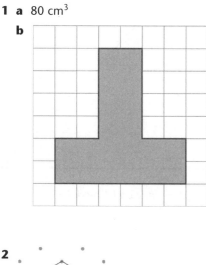

2

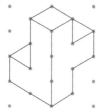

   (1 mark for any isometric view of the five-cube shape)

Page 158 Congruency and tessellations

1 a C; D and F

   b B

2 a A pattern of shapes with no gaps and no overlap

   b

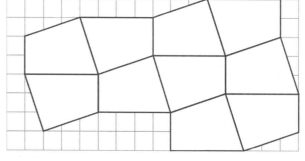

Page 159 Transformations

1 a Translation *(1 mark)* of (–6, –3) *(1 mark)*

   b Triangle at (3, 6), (3, 9), (5, 6)

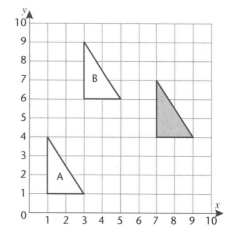

   c (–7, 5)

2

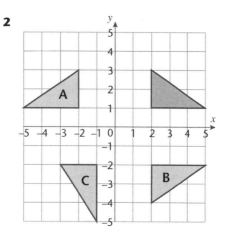

Page 160 Transformations

1 a and **b**

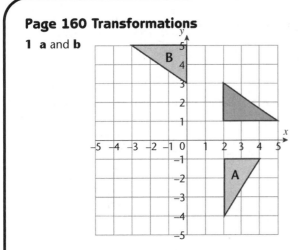

c 90° *(1 mark)* clockwise *(1 mark)* about (4, 3) *(1 mark)*

2 a Enlargement scale factor $\frac{1}{2}$ *(1 mark)*
    centre (1, 8) *(1 mark)*

   b Triangle at (2, 1), (3, 1), (2, $2\frac{1}{2}$)

Page 161 Constructions

1 *Sides 5 cm and 7 cm; included angle 55°*

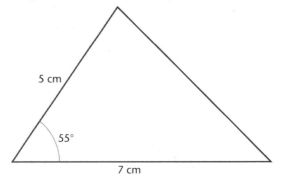

(1 mark for each length ±1 mm and 1 mark for angle ±1°)

2 a

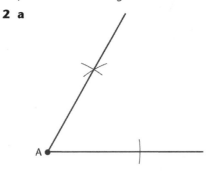

(Deduct a mark if arcs not shown)

   b i *Side 6 cm and 8 cm with included angle 60°*

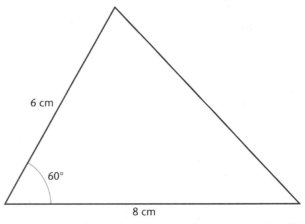

(1 mark for each length ±1 mm and 1 mark for angle ±1°)

   ii 7.2 cm

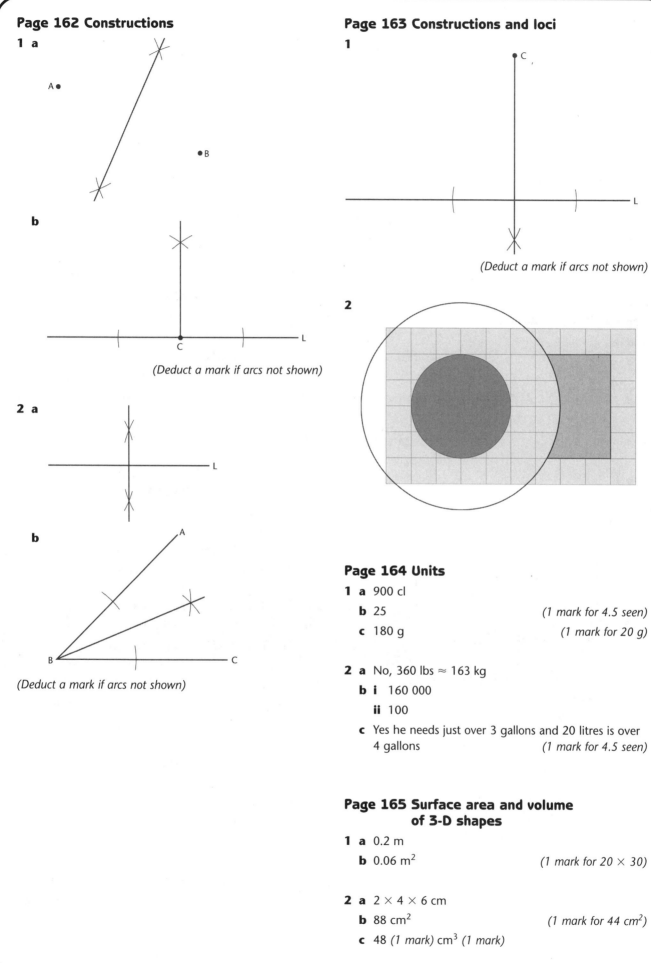

Page 162 Constructions

1 a

b

(Deduct a mark if arcs not shown)

2 a

b

(Deduct a mark if arcs not shown)

Page 163 Constructions and loci

1

(Deduct a mark if arcs not shown)

2

Page 164 Units

1 a 900 cl

b 25 *(1 mark for 4.5 seen)*

c 180 g *(1 mark for 20 g)*

2 a No, 360 lbs ≈ 163 kg

b i 160 000

ii 100

c Yes he needs just over 3 gallons and 20 litres is over 4 gallons *(1 mark for 4.5 seen)*

Page 165 Surface area and volume of 3-D shapes

1 a 0.2 m

b 0.06 m² *(1 mark for 20 × 30)*

2 a 2 × 4 × 6 cm

b 88 cm² *(1 mark for 44 cm²)*

c 48 *(1 mark)* cm³ *(1 mark)*

Page 166 Density and prisms

1 0.0068 kg/cm^3 or 6.8 g/cm^3 *(1 mark for units)*

2 a 14 cm^2
  b 168 cm^3 *(1 mark for 14×12)*
3 160π

Page 167 Pythagoras' theorem

1 12.8 *(1 mark for $\sqrt{164}$)*
2 10.2 *(1 mark for $\sqrt{104}$)*
3 4.5 *(1 mark for $\sqrt{20.41}$)*

Algebra
Page 170 Basic algebra

1 a $x - 2$
  b $2x$
  c $2x - 4$ or $2(x - 2)$
  d $x + x - 2 + 2x + 2x - 4$ *(1 mark)* $= 6x - 6$

2 a $3y \times y = 3y^2$, $2y + y = 3y$, $3(y + 1) = 3y + 3$

3 b i $3q$
    ii $15pq$
    iii $9x - 4$

Page 171 Expanding and factorising

1 a $5x - 15$
  b $8x + 6$
  c $2(2x + 3) + 2(x + 3)$ *(1 mark)* $= 6x + 12$

2 a $3x - 12 + 8x + 2$ *(1 mark)* $= 11x - 10$
  b i $2(2x + 3)$
    ii $x(5x + 2)$

3 a i $5x^2 - 11xy$
    ii $10x - 12y$
  b i $3xy(y + 2x)$
    ii $2ab(2b - 4a + ab)$
  c $3pq(pq + 2)$

Page 172 Quadratic expansion and substitution

1 a $3x + 6$
  b $x^2 + 2x$
  c $x^2 - 3x + 2x - 6$ *(1 mark)* $= x^2 - x - 6$ *(1 mark)*
  d $(x + 2)(x + 1)$ *(1 mark)* $= x^2 + 3x + 2$
2 a $x^2 - 4x + x - 4$ *(1 mark)* $= x^2 - 3x - 4$ *(1 mark)*
  b $(x + 4)(x + 4)$ *(1 mark)* $= x^2 + 8x + 16$
3 a 4
  b 52
  c i $200f + 50e$
    ii £10 000

Page 173 Linear equations

1 a $\frac{x}{3} = -1$ *(1 mark)*; $x = -3$ *(1 mark)*
  b $3x = -3$ *(1 mark)*; $x = -1$ *(1 mark)*
  c $7x = 14$ *(1 mark)*; $x = 2$ *(1 mark)*
2 a $x = 1$ *(1 mark for $2x - x = 6 - 5$)*
  b $y = 5\frac{1}{2}$ *(1 mark for $3y - y = 5 + 6$)*
3 a 8
  b Because $2(1\frac{1}{2} + 5) = 2 \times 6\frac{1}{2} = 13$

Page 174 Linear equations

1 a i $3x + 5 = 26$
    ii 7
  b i $12y = 24$ *(1 mark)*; $y = 2$ *(1 mark)*
    ii $4x = 12$ *(1 mark)*; $x = 3$ *(1 mark)*

2 a $2x = 3$ *(1 mark)*; $x = 1\frac{1}{2}$ *(1 mark)*
  b $2x = -17$ *(1 mark)*; $x = -8\frac{1}{2}$ *(1 mark)*
  c $3x = 18$ *(1 mark)*; $x = 6$ *(1 mark)*

3 a $x = -3$ *(1 mark for $3x = -9$)*
  b $x = -\frac{1}{8}$ *(1 mark for $8x = -1$)*

Page 175 Linear equations

1 a $a = 4$, $b = 6$, $c = 7$, $d = 8$
  b i $50x + 140$
    ii $50x + 140 = 340$ *(1 mark)*, $x = 4$

2 $x + 3x - 1 + 2x + 5 = 25$ *(1 mark)*
   $6x + 4 = 25$ *(1 mark)*
   $x = 3.5$ cm   *(1 mark)*

Page 176 Trial and improvement

1 *1 mark for finding the answer is between 5.6 and 5.7*

1 mark for testing 5.65

1 mark for $x = 5.7$

2 *1 mark for testing 4 (8.5)*

1 mark for finding the answer is between 3.7 and 3.8

1 mark for testing 3.75

1 mark for $x = 3.7$

Page 177 Formulae

1 a $x + 6$

b $4x + 6$

c $4x + 6 = 27$ *(1 mark)*; $x = 5\frac{1}{4}$

2 a It is true for all values

b As for **a**; it is true for all values

3 a $x = \frac{C}{\pi}$

b $3x = 6y + 9$ *(1 mark)*; $x = 2y + 3$ *(1 mark)*

Page 178 Inequalities

1 a $x \leq 2$

b $x > -2$

c $-1, 0, 1, 2$

2 a $-3 < x \leq 1$

b i $\frac{x}{2} > -2$ *(1 mark)*; $x > -4$

ii $x + 3 \leq 2$ *(1 mark)*; $x \leq -1$

c $-3, -2, -1$

Page 179 Graphs

1 a 3 miles

b 13 km

c 100 miles

2 a i 2 km

ii 5 minutes

b i 40 minutes

ii $7\frac{1}{2}$ km/h

Page 180 Linear graphs

1 a 1; 4; 7

b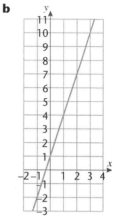

c 2.3

2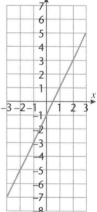

Page 181 Linear graphs

1 a D

b C

c D and E

d 3; $\frac{1}{2}$; $-\frac{5}{3}$

2 Graph of $y = 3x - 2$;
intercepting y-axis at -2 *(1 mark)*;
gradient 3 *(1 mark)*

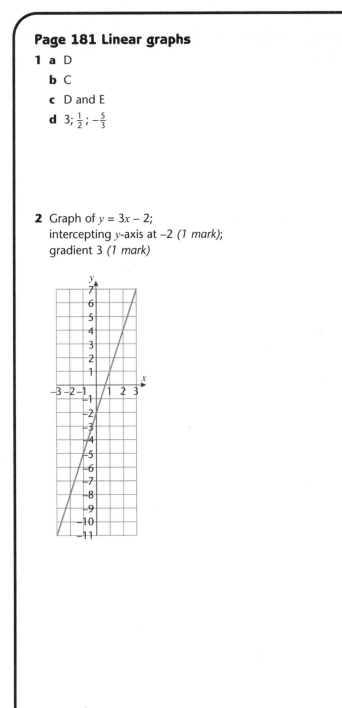

Page 182 Quadratic graphs

1 a 4, 3, 4, 7

b Graph of $y = x^2 + 3$

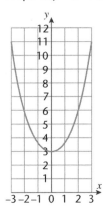

(1 mark for five correct points)

2 a -4, -6, -4

b Graph of $y = x^2 - 3x - 4$

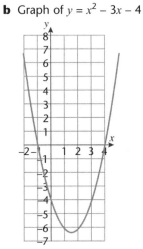

(1 mark for five correct points)

Page 183 Quadratic graphs

1 a 0, 1, 4, 9

  b Graph of $y = x^2 - 2x + 1$

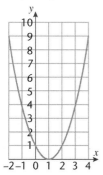

(1 mark for five correct points)

  c −1.45, 3.45

  d 1

2 a 7, −2, −1, 2

  b Graph of $y = x^2 + 2x - 1$

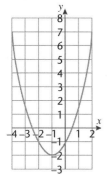

(1 mark for five correct points)

  c −2.9, 0.9

  d −2.4, 0.4

Page 184 Pattern

1 a $1 + 3 + 5 + 7 = 16 = 4^2$; $1 + 3 + 5 + 7 + 9 = 25 = 5^2$

  b 99

  c 225

2 a

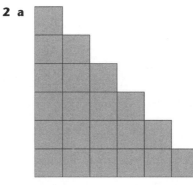

  b 6, 10, 15

  c The number added goes up by 1 more each time

Page 185 The nth term

1 a 5, 9, 13

  b 7th

  c The terms are all odd numbers, 84 is not odd

  d $7n - 4$ *(1 mark each term)*

2 a 16, 21, 26

  b 101

  c $5n + 1$ *(1 mark each term)*

Page 186 Sequences

1 a Always odd

  b Always even

  c Could be either

  d Always odd

  e Always odd

2 a $2 \times$ anything is even

  b $2n \times 2n = 4n^2$
       (1 mark); which is a multiple of 4 *(1 mark)*

3 a 18, 24, 30, 36

  b $6n + 6$ or $6(n + 1)$ *(1 mark each term or factor)*

Notes